LONDON MATHEMATICAL SOCIETY LECTURE NOTE SERIES

Managing Editor: Professor I.M. James.
Mathematical Institute, 24-29 St G

1. General cohomology theory and I
4. Algebraic topology, J.F.ADAMS
5. Commutative algebra, J.T.KNIGHT
8. Integration and harmonic analysi ...ups, R.E.EDWARDS
9. Elliptic functions and elliptic ... .DU VAL
10. Numerical ranges II, F.F.BONSALL & J.DUNCAN
11. New developments in topology, G.SEGAL (ed.)
12. Symposium on complex analysis, Canterbury, 1973, J.CLUNIE & W.K.HAYMAN (eds.)
13. Combinatorics: Proceedings of the British Combinatorial Conference 1973, T.P.McDONOUGH & V.C.MAVRON (eds.)
15. An introduction to topological groups, P.J.HIGGINS
16. Topics in finite groups, T.M.GAGEN
17. Differential germs and catastrophes, Th.BROCKER & L.LANDER
18. A geometric approach to homology theory, S.BUONCRISTIANO, C.P. BOURKE & B.J.SANDERSON
20. Sheaf theory, B.R.TENNISON
21. Automatic continuity of linear operators, A.M.SINCLAIR
23. Parallelisms of complete designs, P.J.CAMERON
24. The topology of Stiefel manifolds, I.M.JAMES
25. Lie groups and compact groups, J.F.PRICE
26. Transformation groups: Proceedings of the conference in the University of Newcastle-upon-Tyne, August 1976, C.KOSNIOWSKI
27. Skew field constructions, P.M.COHN
28. Brownian motion, Hardy spaces and bounded mean oscillations, K.E.PETERSEN
29. Pontryagin duality and the structure of locally compact Abelian groups, S.A.MORRIS
30. Interaction models, N.L.BIGGS
31. Continuous crossed products and type III von Neumann algebras, A.VAN DAELE
32. Uniform algebras and Jensen measures, T.W.GAMELIN
33. Permutation groups and combinatorial structures, N.L.BIGGS & A.T.WHITE
34. Representation theory of Lie groups, M.F. ATIYAH et al.
35. Trace ideals and their applications, B.SIMON
36. Homological group theory, C.T.C.WALL (ed.)
37. Partially ordered rings and semi-algebraic geometry, G.W.BRUMFIEL
38. Surveys in combinatorics, B.BOLLOBAS (ed.)
39. Affine sets and affine groups, D.G.NORTHCOTT
40. Introduction to Hp spaces, P.J.KOOSIS
41. Theory and applications of Hopf bifurcation, B.D.HASSARD, N.D.KAZARINOFF & Y-H.WAN
42. Topics in the theory of group presentations, D.L.JOHNSON
43. Graphs, codes and designs, P.J.CAMERON & J.H.VAN LINT
44. Z/2-homotopy theory, M.C.CRABB
45. Recursion theory: its generalisations and applications, F.R.DRAKE & S.S.WAINER (eds.)
46. p-adic analysis: a short course on recent work, N.KOBLITZ
47. Coding the Universe, A.BELLER, R.JENSEN & P.WELCH
48. Low-dimensional topology, R.BROWN & T.L.THICKSTUN (eds.)

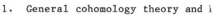

49. Finite geometries and designs, P.CAMERON, J.W.P.HIRSCHFELD & D.R.HUGHES (eds.)
50. Commutator calculus and groups of homotopy classes, H.J.BAUES
51. Synthetic differential geometry, A.KOCK
52. Combinatorics, H.N.V.TEMPERLEY (ed.)
53. Singularity theory, V.I.ARNOLD
54. Markov processes and related problems of analysis, E.B.DYNKIN
55. Ordered permutation groups, A.M.W.GLASS
56. Journées arithmétiques 1980, J.V.ARMITAGE (ed.)
57. Techniques of geometric topology, R.A.FENN
58. Singularities of smooth functions and maps, J.MARTINET
59. Applicable differential geometry, M.CRAMPIN & F.A.E.PIRANI
60. Integrable systems, S.P.NOVIKOV et al.
61. The core model, A.DODD
62. Economics for mathematicians, J.W.S.CASSELS
63. Continuous semigroups in Banach algebras, A.M.SINCLAIR
64. Basic concepts of enriched category theory, G.M.KELLY
65. Several complex variables and complex manifolds I, M.J.FIELD
66. Several complex variables and complex manifolds II, M.J.FIELD
67. Classification problems in ergodic theory, W.PARRY & S.TUNCEL
68. Complex algebraic surfaces, A.BEAUVILLE
69. Representation theory, I.M.GELFAND et. al.
70. Stochastic differential equations on manifolds, K.D.ELWORTHY
71. Groups - St Andrews 1981, C.M.CAMPBELL & E.F.ROBERTSON (eds.)
72. Commutative algebra: Durham 1981, R.Y.SHARP (ed.)
73. Riemann surfaces: a view toward several complex variables, A.T.HUCKLEBERRY
74. Symmetric designs: an algebraic approach, E.S.LANDER
75. New geometric splittings of classical knots (algebraic knots), L.SIEBENMANN & F.BONAHON
76. Linear differential operators, H.O.CORDES
77. Isolated singular points on complete intersections, E.J.N.LOOIJENGA
78. A primer on Riemann surfaces, A.F.BEARDON
79. Probability, statistics and analysis, J.F.C.KINGMAN & G.E.H.REUTER (eds.
80. Introduction to the representation theory of compact and locally compact groups, A.ROBERT
81. Skew fields, P.K.DRAXL
82. Surveys in combinatorics: Invited papers for the 9th British Combinatorial Conference 1983, E.K.LLOYD (ed.)
83. Homogeneous structures on Riemannian manifolds, F.TRICERRI & L.VANHECKE

London Mathematical Society Lecture Notes Series. 83

Homogeneous Structures on Riemannian Manifolds

F. TRICERRI
Department of Mathematics, Polytechnic of Turin

L. VANHECKE
Department of Mathematics, Katholieke Universiteit, Leuven

CAMBRIDGE UNIVERSITY PRESS
Cambridge
London New York New Rochelle
Melbourne Sydney

CAMBRIDGE UNIVERSITY PRESS
Cambridge, New York, Melbourne, Madrid, Cape Town, Singapore, São Paulo

Cambridge University Press
The Edinburgh Building, Cambridge CB2 8RU, UK

Published in the United States of America by Cambridge University Press, New York

www.cambridge.org
Information on this title: www.cambridge.org/9780521274890

First published 1983
Re-issued in this digitally printed version 2008

A catalogue record for this publication is available from the British Library

Library of Congress Catalogue Card Number: 83–2097

ISBN 978-0-521-27489-0 paperback

To Magda and Nuccia

CONTENTS

PREFACE

It is a pleasure for us to thank the Department of Mathematics of the University of Durham, the Katholieke Universiteit Leuven and the Politecnico di Torino for their hospitality during our research.

We also thank the C.N.R. (Italy) and the N.F.W.O. (Belgium) for their financial support.

There were many people who were involved in one way or another with our research and with the writing of these lecture notes. In particular, we are grateful to L. Gheysens, A. Kaplan, O. Kowalski, F. Ricci, A. Sanini, P. Verheyen and T.J. Willmore for several useful discussions, for the information they provided, for their encouragement and for their friendship.

Our thanks to Bea Peeters for typing the manuscript.

Leuven and Torino, August 1982

- Dipartimento di Matematica, Politecnico di Torino, Corso Duca degli Abruzzi 24, 10129 Torino, Italy.

- Departement Wiskunde, Katholieke Universiteit Leuven, Celestijnenlaan 200 B, B-3030 Leuven, Belgium.

INTRODUCTION

As is well-known, E. Cartan proved that a connected, complete and simply connected Riemannian manifold is a symmetric space if and only if the curvature is constant under parallel translation. In 1958 Ambrose and Singer [1] extended this theory and gave a characterization of *homogeneous Riemannian manifolds* by a local condition which is to be satisfied at all points. More specifically, they proved that a connected, complete and simply connected Riemannian manifold (M,g) is homogeneous, i.e. there exists a transitive and effective group G of isometries of M, if and only if there exists a tensor field T of type (1,2) such that

$$
\text{(AS)} \qquad \begin{cases} \text{(i)} & g(T_X Y, Z) + g(Y, T_X Z) = 0 \,, \\ \text{(ii)} & (\nabla_X R)_{YZ} = [T_X, R_{YZ}] - R_{T_X YZ} - R_{YT_X Z} \,, \\ \text{(iii)} & (\nabla_X T)_Y = [T_X, T_Y] - T_{T_X Y} \,, \end{cases}
$$

for $X,Y,Z \in \mathfrak{X}(M)$. Here ∇ denotes the Levi Civita connection and R is the Riemann curvature tensor of M.

These conditions are also used to study weakly locally homogeneous, infinitesimally homogeneous and curvature homogeneous manifolds [1],[44].

Although there are shorter proofs than that given in [1] (see for example [28]) the treatment of Ambrose and Singer has the advantage of being *constructive*. Indeed, the authors give an explicit construction of the tensor T when the group G is given and conversely, they determine the group G when a particular T is given. In chapter 1 of these notes we give a short proof of the theorem of Ambrose and Singer but we also give a full version of the constructive proof because on the one hand a lot of geometers are not familiar with it and on the other hand some important results are needed in the rest of these notes.

The proof of Ambrose and Singer sets up a natural correspondence between groups G of which M is a homogeneous Riemannian manifold and the T's which satisfy the conditions (AS). In that paper the authors make the following remark : "This suggests the possibility of classifying the groups G of which M is a Riemannian homogeneous space thru the T's. It also suggests the possibility of classifying Riemannian homogeneous manifolds by properties of the T's." They give two examples concerning this last suggestion. We will come back on these examples later on.

Our interest in this theorem and these remarks arose during our research on *harmonic spaces* and *spaces with volume-preserving geodesic symmetries* (see [51]). First of all we wanted to have a method to be able to decide whether a harmonic manifold is a homogeneous space or not. Secondly we tried to find manifolds with volume-preserving geodesic symmetries which are not naturally reductive. To decide whether a homogeneous manifold is naturally reductive is far from easy in many cases and therefore we were looking for a characterization using the tensors T instead of using all the groups of isometries of the Riemannian manifold. At the same time we wanted to have a method which was closely related to the curvature and the geometric or Riemannian properties of the manifold.

The first suggestion made by Ambrose and Singer gives rise to a difficult problem and this for several reasons. Let (M,g) be a homogeneous Riemannian space with a given group G of isometries. Following the method of Ambrose and Singer, this determines a tensor field T. Now using this T one can determine conversely the group of isometries and this is in general a group G' which is not isomorphic to G. The Euclidean plane $\mathbb{R}^2$ provides a simple example. Let G denote the group of all isometries of $\mathbb{R}^2$. Then one obtains T = 0. Further, the construction of G' starting from T = 0 gives for G' only the group of translations of $\mathbb{R}^2$. Hence an important problem will be to understand for which spaces we have G = G'. Only for this case the solutions of the equations (AS) will give a parametrization of the transitive and effective groups G. This means also that we have to understand which groups can be obtained from the solutions of the equations (AS) and how we can characterize these groups. We refer to chapter 1 and chapter 2 for more detailed information about this problem.

Concerning the second suggestion they consider the example of homogeneous Riemannian manifolds such that for all $X,Y \in \mathfrak{X}(M)$ we have

$$T_X Y + T_Y X = 0.$$

Moreover, they prove that for these spaces the geodesics of the Riemannian connection are orbits of one-parameter subgroups of G. In chapter 9 we discuss a remarkable example of A. Kaplan which shows that the converse property does not hold. In fact we will prove that the condition for T characterizes naturally reductive homogeneous spaces. This implies that Theorem 5.4 in [1] has to be modified (see chapter 6).

This condition on the T is an *algebraic* condition which is invariant under the action of the orthogonal group. For this reason we study the decomposition of the space of tensors T which satisfy the condition ((AS)(i)) into irreducible factors under the action of the orthogonal group. In this way we obtain a set of algebraic conditions for the tensor T. These conditions are invariant under the action of the orthogonal group and they provide a kind of *classification* for the homogeneous Riemannian spaces into eight different classes. This method is similar to that used in [14] for the study of Einstein-like manifolds and in [15] to give a classification of almost Hermitian manifolds into sixteen classes. See [49] for a similar treatment of the space of curvature tensors on an almost Hermitian manifold but under the action of the unitary group. See also [6] for the orthogonal group.

In chapter 2 we first define and treat *homogeneous Riemannian structures* from a general viewpoint. Such a structure is given on a Riemannian manifold by a solution of the equations (AS). Note that a solution of the equations (AS), in general, is not uniquely determined. For example, let T be a homogeneous Riemannian structure on (M,g) and φ an isometry of M. Then the tensor T' given by

$$T'_X Y = \varphi_* T_{\varphi_*^{-1} X} \varphi_*^{-1} Y ,$$

$X,Y \in \mathfrak{X}(M)$, determines also a homogeneous Riemannian structure and, in general, T' is different from T. This leads to the definition of *isomorphic homogeneous structures*. But it also can happen that on the same manifold (M,g) there exist two *nonisomorphic* homogeneous structures T and T'. The point is that T and T' give rise to two nonisomorphic transitive groups of isometries or to the same group but with different reductive decompositions of the Lie algebra of the group. We refer to chapters 7 and 8 for detailed examples of these possibilities.

Chapter 3 contains the *algebraic* part. Here we give the decomposition mentioned earlier, we determine the quadratic invariants and we write down the projections of T on the irreducible factors.

In chapter 4 we concentrate on the homogeneous Riemannian structures on *two-dimensional manifolds* and give the complete classification. More specifically, we show that the Poincaré half-plane is the only connected, complete and simply connected surface which has a homogeneous Riemannian structure T *different from zero*. Moreover, up to isomorphism, this nonvanishing structure is unique.

Next, in chapter 5, we study the class $\mathcal{T}_1$ of homogeneous structures with defining condition

$$T_X Y = g(X,Y)\xi - g(Y,\xi)X ,$$

where ξ is a given vector field on (M,g) and $X,Y \in \mathfrak{X}(M)$. This is the immediate analogue of the case for surfaces since for algebraic reasons all the homogeneous structures on surfaces are of this type. We prove that the *hyperbolic space* is the only space (connected, complete, simply connected) having such a structure $T \neq 0$.

The main result of chapter 6 is the characterization of *naturally reductive homogeneous* spaces by the condition "$T_X X = 0$ for all $X \in \mathfrak{X}(M)$". These are by far the simplest kind of homogeneous spaces. There are many examples known. For example, all the homogeneous Riemannian spaces whose isotropy representation is irreducible (in particular, the irreducible symmetric spaces) belong to this class. See the classification given by J. Wolf [57]. See also [12] and chapters 6,7 and 8. Further we give a complete list of all the *three-dimensional* connected, complete and simply connected manifolds which admit such a nonvanishing structure. The main point is that this is done here with the help of the curvature tensor R of the manifold and the tensor T, concentrating in this way on the Riemannian viewpoint.

In chapter 7 and chapter 8 we treat in detail some important examples. Chapter 7 is completely devoted to the study of the *Heisenberg group*. In chapter 8 we consider all the *Lie groups of dimension three*, the 3-*symmetric spaces*, the *four-dimensional hyperbolic space* and some other *four-dimensional Lie groups*. On the one hand these examples illustrate several general statements and theorems and on the other hand they provide examples for the eight classes and their respective inclusions.

Motivated by our research on manifolds with volume-preserving geodesic symmetry (see [51]) A. Kaplan discovered a nice six-dimensional example with this property but which is not naturally reductive. The same example shows that there are manifolds such that all geodesics are orbits of one-parameter subgroups of isometries which are not naturally reductive. This example is a *group of type* H or a *generalized Heisenberg group*. In chapter 9 we give a brief survey on these groups and prove several of Kaplan's results. We do this in a different way using the methods of these notes. Finally we show that the six-dimensional example has some new additional properties for the geodesic symmetries which are much stronger than the volume-preserving property. These properties are also valid for naturally reductive homogeneous spaces and hence, the example shows again that the given properties are not characteristic for naturally reductive spaces.

As is now well-known, the decomposition of the space of curvature tensors on a Riemannian manifold has three irreducible invariant components under the action of the orthogonal group. When one considers four-dimensional manifolds and the special orthogonal group, then one of the components splits further into two irreducible spaces of the same dimension. This gives rise to the notion of self-dual and anti-self-dual curvature tensors. In chapter 10 we consider the space of tensors T as before but now under the action of the special orthogonal group and for dimension four. We show that also in this case one of the irreducible factors for O(4) splits into two irreducible spaces of the same dimension for SO(4). This leads to the definition of *self-dual* and *anti-self-dual homogeneous structures*. We provide examples of such structures.

There are several problems which need further research. We mention the following ones here. In the first place it is shown in chapter 6 how the manifolds with a structure of type $\mathcal{T}_3$, i.e. a naturally reductive structure, can be characterized by means of properties of the geodesics or geometrical notions related to the geodesics. It would be interesting to have similar characterizations for all the other classes. Secondly, further research for examples is needed to consider the inclusion relations for the sixteen classes in chapter 10 and to lend more substance to the introduction of self-dual and anti-self-dual homogeneous structures.

The bibliography has been kept to a minimum and consists only of papers and books referred to in these notes. Further information, and a more complete list of papers on homogeneous manifolds, can be found in the books and papers cited.

1. THE THEOREM OF AMBROSE AND SINGER

The main purpose of this section is to give a proof of the theorem of Ambrose and Singer and to concentrate on several facts we need later on. There are three sections. In section A we collect some preliminary material used in the two other sections and in the rest of these notes. In section B we give a more direct proof of the theorem, inspired by the method used in [27],[28] in a more general context. (See also [37],[38].) Finally in section C we consider the proof given by Ambrose and Singer with slight modifications. More specifically, we give the explicit construction of the transitive group G of isometries and its algebra $\mathfrak{g}$ when the tensor field T is given. This is a fundamental construction for the development of the theory of homogeneous structures. At many places we refer to [26] as a standard reference for a lot of notions, theorems and additional material.

A. PRELIMINARIES

Let (M,g) be an n-dimensional Riemannian manifold of class C^∞. Let ∇ denote the Levi Civita connection of (M,g) and R the corresponding Riemann curvature tensor given by

$$R_{XY} = \nabla_{[X,Y]} - [\nabla_X, \nabla_Y] , \qquad X,Y \in \mathfrak{X}(M) , \tag{1.1}$$

where $\mathfrak{X}(M)$ is the Lie algebra of C^∞ vector fields on M.

Let φ be an *isometry* of (M,g). Then φ is also an *affine transformation* of ∇. This means :

$$\varphi_*(\nabla_X Y) = \nabla_{\varphi_* X}\varphi_* Y \tag{1.2}$$

for $X,Y \in \mathfrak{X}(M)$. Here φ_* denotes the differential of φ.

A vector field ξ on M is called an *infinitesimal isometry* or a *Killing vector field* if the local one-parameter group of local transformations generated by ξ in a neighbourhood of each point of M consists of local isometries. This is equivalent to the following condition :

$$(\mathcal{L}_\xi g)(X,Y) = \xi g(X,Y) - g([\xi,X],Y) - g(X,[\xi,Y]) = 0 , \tag{1.3}$$

$X,Y \in \mathfrak{X}(M)$. $\mathcal{L}_\xi$ denotes the *Lie derivative* with respect to ξ. Now put

$$A_X Y = -\nabla_Y X = \mathcal{L}_X Y - \nabla_X Y. \tag{1.4}$$

Then (1.3) becomes

$$g(A_\xi X,Y) + g(X,A_\xi Y) = 0. \tag{1.5}$$

Hence ξ is a Killing vector field if and only if A_ξ is *skew-symmetric*.

It is well-known (see for example [26, vol. I, p. 239]) that the *group* $\mathfrak{J}(M)$ *of all isometries* of a Riemannian manifold is a Lie group of transformations of M. The Lie algebra i(M) of $\mathfrak{J}(M)$ is isomorphic to the Lie algebra of *complete Killing vector fields*.

Next we recall

PROPOSITION 1.1. *Let* (M,g) *be a connected Riemannian manifold and* φ,ψ *two isometries of* M. *Suppose there exists a point* p *of* M *such that*

$$\varphi(p) = \psi(p) , \quad \varphi_*(p) = \psi_*(p).$$

Then $\varphi = \psi$.

For a proof see [19, p. 62]. It follows that an isometry of a connected Riemannian manifold (M,g) is completely determined by its value and its differential at a single point. The following proposition gives an infinitesimal version of this result.

PROPOSITION 1.2. *Let* ζ *and* ξ *be two Killing vector fields on a connected Riemannian manifold* M. *Suppose*

$$\zeta|_p = \xi|_p \quad , \quad A_\zeta|_p = A_\xi|_p$$

for some $p \in M$. *Then* $\zeta = \xi$.

We refer to [27] for a proof.

Among the affine transformations of (M,∇) the isometries of (M,g) are characterized as follows.

PROPOSITION 1.3. *Let* (M,g) *be a connected Riemannian manifold and let* φ *be an affine transformation with respect to the Levi Civita connection* ∇. *Suppose there exists a point* p *of* M *where* $\varphi_*(p)$ *is an isometry. Then* φ *is an isometry.*

Note that the fact that the *parallel transport* with respect to ∇ along a curve is an isometry is the important point in the proof (see [19, p. 201]). This is still true if ∇ is replaced by an arbitrary metric connection $\tilde{\nabla}$, i.e. $\tilde{\nabla}g = 0$. Hence Proposition 1.3 is also valid if one replaces ∇ by $\tilde{\nabla}$.

We will also need the following proposition. We recall briefly the proof of [1].

PROPOSITION 1.4. *Let* (M,g) *be a complete Riemannian manifold. If* X *is a vector field such that its norm* $\|X\|^2 = g(X,X)$ *is bounded, then* X *is a complete vector field.*

Proof. Let $\|X\| \leqslant k$ where k is a constant. Further, let $\varphi(t)$ be an integral curve of X defined for $a < t < b$. Now let $\{t_n\}$ be an infinite sequence converging to b. Then the sequence $\varphi(t_n)$ is a Cauchy sequence. Indeed, the distance between $\varphi(t_n)$ and $\varphi(t_m)$ is such that

$$d(\varphi(t_n),\varphi(t_m)) \leqslant \int_{t_n}^{t_m} \|\dot{\varphi}(t)\|\,dt \leqslant k(t_m - t_n).$$

Since (M,g) is complete, $\varphi(t_n)$ converges to a point p of M. It is clear that p is independent of the sequence $\{t_n\}$ chosen before.

Next choose the integral curve $\psi(s)$ of X through p and take

$\psi(0) = p$. This integral curve is defined on an interval $(-\varepsilon,\varepsilon)$. Now put

$$\tilde{\varphi}(t) = \begin{cases} \varphi(t) & \text{for } a < t < b , \\ \psi(t-b) & \text{for } b - \varepsilon \leqslant t < b + \varepsilon. \end{cases}$$

Then $\tilde{\varphi}(t)$ is an integral curve of X which is defined for all t in the interval $(a,b+\varepsilon)$. This implies that each *maximal* integral curve of X is necessarily defined for all $t \in \mathbb{R}$. Hence X is complete.

Concerning *complete* Riemannian manifolds we now give a proof of the following fact :

PROPOSITION 1.5. *Let* (M,g) *be a complete Riemannian manifold. Then each metric connection* $\tilde{\nabla}$ *on* M *is complete.*

Proof. This result is classical in the case when the metric connection $\tilde{\nabla}$ is the Levi Civita connection of (M,g) (see [26, vol. I, p. 172] or [41, p. 102]).

Now let $\tilde{\nabla}$ be an arbitrary metric connection and let $\gamma(t)$ be a geodesic of $\tilde{\nabla}$ defined for $a < t < b$. Choose an infinite sequence $\{t_n\}$ which has b as limit. Since $\tilde{\nabla}$ is metric, $\|\dot{\gamma}(t)\| = k$ where k is constant. It is easy to see that $\{\gamma(t_n)\}$ is a Cauchy sequence and hence it converges to a point p of M.

Let $\mathcal{U}$ be the (relatively compact) domain of a system of normal coordinates $(x^1,\ldots,x^n)$ centered at p. Then the functions $\gamma^i(t) = x^i(\gamma(t))$, $i = 1,\ldots,n$, are defined for $c < t < b$ where $a < c < b$. Since $\|\dot{\gamma}(t)\| = k$ is constant, the functions $\dot{\gamma}^i(t)$ are bounded. Hence, the functions $\ddot{\gamma}^i(t)$ are also bounded since

$$\ddot{\gamma}^i + \sum_{j,k} \tilde{\Gamma}^i_{jk}\dot{\gamma}^j\dot{\gamma}^k = 0.$$

Here the $\tilde{\Gamma}^i_{jk}$ are the local components of the connection $\tilde{\nabla}$. It follows from the mean value theorem that the $\dot{\gamma}^i(t)$ are uniformly continuous and this implies that $\lim \dot{\gamma}^i(t)$ exists when $t \to b$. Put

$$\lim_{t\to b} \dot{\gamma}^i(t) = a^i \quad , \quad i = 1,\ldots,n$$

and let $u \in T_pM$ be the vector

$$u = \sum_{i=1}^{n} a^i \frac{\partial}{\partial x_i}\Big|_p .$$

The geodesic of $\tilde{\nabla}$ which is tangent to u at p is given by

$$\psi(s) = (a^1 s,\dots,a^n s) \ , \quad -\varepsilon < s < \varepsilon.$$

Hence

$$\tilde{\gamma}(t) = \begin{cases} \gamma(t) & \text{for } a < t < b \ , \\ \psi(t-b) & \text{for } b \leqslant t < b + \varepsilon \end{cases}$$

is a curve of class C^2. Since $\tilde{\gamma}(t)$ satisfies the system of equations for a geodesic, it is an extension of $\gamma(t)$. This implies that $\tilde{\nabla}$ is complete.

In what follows we begin by concentrating on the *principal fibre bundle* $\mathcal{O}(M)$ of *orthonormal frames* of (M,g) and on some of its properties. The structure group of $\mathcal{O}(M)$ is the orthogonal group $O(n)$, where n is the dimension of M. A point u of $\mathcal{O}(M)$ is a pair $(p;u_1,\dots,u_n)$ where $p \in M$ and $(u_1,\dots,u_n)$ is an orthonormal frame of T_pM. The projection $\pi : \mathcal{O}(M) \to M$ is determined by $\pi(u) = p$. Further, let $\mathcal{U}$ be an open neighbourhood of $p = \pi(u)$ and let $(E_1,\dots,E_n)$ be a local orthonormal frame field on $\mathcal{U}$ (a *local cross section* of $\mathcal{O}(M)$). Then, for all $v = (q;v_1,\dots,v_n)$ of $\pi^{-1}(\mathcal{U})$, we have

$$v_h = \sum_m a_h^m E_m|_q \ , \tag{1.6}$$

where $a = (a_h^m)$ is an element of $O(n)$. Hence the map $v \mapsto (\pi(v),a)$ is a diffeomorphism of $\pi^{-1}(\mathcal{U})$ onto $\mathcal{U} \times O(n)$. So we may identify $\pi^{-1}(\mathcal{U})$ with $\mathcal{U} \times O(n)$ and the tangent space of $\mathcal{O}(M)$ at $v \in \pi^{-1}(\mathcal{U})$ can be identified with the direct sum

$$T_{\pi(v)}M \oplus T_aO(n).$$

Hence a tangent vector $\bar{X}$ of $T_v\mathcal{O}(M)$ can be expressed as

$$\bar{X} = X + aA , \tag{1.7}$$

where $X = \pi_*(\bar{X})$ and $A \in \mathfrak{so}(n)$. Here $\mathfrak{so}(n)$ is the Lie algebra of $O(n)$, identified as usual with the tangent space of $O(n)$ at the identity.

Recall that $O(n)$ acts *freely* (without fixed points) on $\mathcal{O}(M)$ and *transitively* (on the right) on the fibres $\pi^{-1}(p)$. The action of $O(n)$ is given by

$$(p;u_1,\ldots,u_n)b = (p; \sum_m b_1^m u_m,\ldots, \sum_m b_n^m u_m) \tag{1.8}$$

where $b = (b_k^h) \in O(n)$. Identifying $\pi^{-1}(\mathcal{U})$ with $\mathcal{U} \times O(n)$, we can write

$$(p,a)b = (p,ab). \tag{1.9}$$

Next let

$$u(t) = (p(t);u_1(t),\ldots,u_n(t))$$

be a curve of $\mathcal{O}(M)$. One says that $u(t)$ is a *horizontal curve with respect to a metric connection* $\tilde{\nabla}$ if all the vector fields $u_h(t)$, $1 \leqslant h \leqslant n$, are *parallel* along the curve $p(t)$ of M. (See for example [26, vol. I].) Hence $u(t)$ is horizontal if and only if, locally, we have

$$\dot{a}_k^h(t) + \sum_m \tilde{\omega}_m^h(\dot{\gamma}(t))a_k^m(t) = 0. \tag{1.10}$$

The $\tilde{\omega}_j^i$ are the local forms of the connection $\tilde{\nabla}$, i.e.

$$\tilde{\omega}_j^i(X) = g(\tilde{\nabla}_X E_j, E_i) \tag{1.11}$$

where $(E_1,\ldots,E_n)$ is the local section of $\mathcal{O}(M)$ which gives the identification of $\pi^{-1}(\mathcal{U})$ with $\mathcal{U} \times O(n)$.

A vector $\bar{X}$ of $T_u\mathcal{O}(M)$ is said to be *horizontal* if it is a tangent vector of a horizontal curve. Hence it follows from (1.7) and

(1.10) that

$$\bar{X} = X + aA \tag{1.12}$$

is horizontal if and only if

$$X = \pi_*(\bar{X}) , \tag{1.13}$$

$$A = -a^{-1}\tilde{\omega}(X)a \tag{1.14}$$

where $\tilde{\omega}(X)$ is the matrix $(\tilde{\omega}^i_j(X)) \in so(n)$.

The horizontal vectors generate a subspace H_u of $T_u\mathcal{O}(M)$, called the *horizontal space*. The *vertical space* is the subspace of $T_u\mathcal{O}(M)$ which is tangent to the fibre through u. We recall (see [26, vol. I]) that the map $u \mapsto H_u$ determines an *infinitesimal connection* on $\mathcal{O}(M)$, *associated to* $\tilde{\nabla}$, i.e. we have

i) H_u depends differentiably on u;

ii) $H_u \oplus V_u = T_u\mathcal{O}(M)$;

iii) $(R_b)_* H_u = H_{ub}$,

where $(R_b)_*$ denotes the differential of the right translation with $b \in O(n)$.

Now let $\xi = (\xi^1,\ldots,\xi^n)$ be an element of $\mathbb{R}^n$. Each u of $\mathcal{O}(M)$ defines an isomorphism between $\mathbb{R}^n$ and T_pM, $p = \pi(u)$, as follows :

$$u(\xi) = \xi^1 u_1 + \ldots + \xi^n u_n .$$

The *standard horizontal vector field* corresponding to ξ and with respect to $\tilde{\nabla}$ is the vector field $B(\xi)$ such that $B(\xi)_{|u}$ is the unique horizontal vector with

$$\pi_*(B(\xi)_{|u}) = u(\xi).$$

Further, let $A \in so(n)$. A^* denotes the *fundamental vector field* corresponding to A, i.e. it is the vertical vector field generated by the one-parameter group of transformations of $\mathcal{O}(M)$ determined by $u \mapsto u(\exp tA)$.

Note that (see [26, vol. I, p. 51 and p. 119])

$$(R_a)_* A^* = (ad(a^{-1})A)^* , \qquad a \in O(n) , \tag{1.15}$$

$$(R_a)_* B(\xi) = B(a^{-1}\xi) , \quad a \in O(n), \ \xi \in \mathbb{R}^n. \tag{1.16}$$

The standard horizontal vector fields generate the horizontal distribution and the fundamental vector fields generate the vertical distribution. Hence we can define a *Riemannian metric* $g_{\tilde{\nabla}}$ on $\mathcal{O}(M)$, depending on $\tilde{\nabla}$, by putting

$$g_{\tilde{\nabla}}(B(\xi),B(\eta)) = \langle \xi,\eta \rangle , \quad \xi,\eta \in \mathbb{R}^n , \tag{1.17}$$

$$g_{\tilde{\nabla}}(A_1^*,A_2^*) = -tr(A_1A_2) , \quad A_1,A_2 \in \mathfrak{so}(n) , \tag{1.18}$$

$$g_{\tilde{\nabla}}(B(\xi),A^*) = 0 , \qquad \xi \in \mathbb{R}^n, \ A \in \mathfrak{so}(n). \tag{1.19}$$

$\langle \ , \ \rangle$ denotes the inner product on $\mathbb{R}^n$.

It is clear that the projection $\pi : \mathcal{O}(M) \to M$ is a *Riemannian submersion*. Hence if $\tilde{d}$ denotes the distance function of $(\mathcal{O}(M),g_{\tilde{\nabla}})$ and d the distance function of (M,g) respectively, we have

$$d(\pi(u),\pi(v)) \leq \tilde{d}(u,v). \tag{1.20}$$

This remark is important for the following proposition.

<u>PROPOSITION 1.6</u>. *Let* (M,g) *be complete. Then* $(\mathcal{O}(M),g_{\tilde{\nabla}})$ *is also complete.*

<u>Proof</u>. To prove that $(\mathcal{O}(M),g_{\tilde{\nabla}})$ is complete, it is sufficient to show that the closure $\bar{A}$ of a bounded subset A of $\mathcal{O}(M)$ is compact (see [26, vol. I, p. 172]). It follows from (1.20) that if A is bounded, then $\pi(A)$ is also bounded. Hence $\overline{\pi(A)}$ is compact. Further, since the fibre of $\mathcal{O}(M)$ is compact, $\pi^{-1}(\overline{\pi(A)})$ is compact ([47, p. 13]) and closed. This implies that $\bar{A} \subset \pi^{-1}(\overline{\pi(A)})$ and hence, since $\bar{A}$ is closed, it is compact.

REMARK. Proposition 1.5 also follows from Proposition 1.4 and Proposition 1.6. Indeed, $\tilde{\nabla}$ is complete if and only if any standard horizontal vector field $B(\xi)$ with respect to $\tilde{\nabla}$ is complete (see [26, vol. I, p. 140]). But since

$$\| B(\xi) \| = \| \xi \|_{\mathbb{R}^n} ,$$

$\| B(\xi) \|$ is bounded. This implies the result.

The definition of $g_{\tilde{\nabla}}$ and (1.15), (1.16) imply

PROPOSITION 1.7. *All the right translations* R_a, $a \in O(n)$, *act as isometries on* $(\mathcal{O}(M), g_{\tilde{\nabla}})$.

Next, let $\tilde{\nabla}$ be the Levi Civita connection ∇ of (M,g). Then we have

PROPOSITION 1.8. *Each isometry* φ *of* (M,g) *induces an isometry* $\tilde{\varphi}$ *of* $(\mathcal{O}(M), g_{\nabla})$ *by*

$$\tilde{\varphi}(u) = \tilde{\varphi}(p;u_1,\dots,u_n) = (\varphi(p);\varphi_*(u_1),\dots,\varphi_*(u_n)). \tag{1.21}$$

Proof. First we note that

$$\tilde{\varphi}(ua) = \tilde{\varphi}(u)a , \qquad u \in \mathcal{O}(M),\ a \in O(n).$$

Hence if $A \in \mathfrak{so}(n)$, we have

$$\tilde{\varphi}_{*|u}(A^*_{|u}) = A^*_{|\tilde{\varphi}(u)} . \tag{1.22}$$

Since φ is an isometry, it preserves the parallelism. So if X is a horizontal vector field of $\mathcal{O}(M)$, $\tilde{\varphi}_*(X)$ is also horizontal.

Next we have

$$\pi \circ \tilde{\varphi} = \varphi \circ \pi$$

and hence

$$(\pi \circ \tilde{\varphi})_{*|u}(B(\xi)_{|u}) = (\varphi \circ \pi)_{*|u}(B(\xi)_{|u})$$

$$= \varphi_{*|p}u(\xi) = \tilde{\varphi}(u)(\xi).$$

The uniqueness of $B(\xi)$ leads to

$$\tilde{\varphi}_{*|u}(B(\xi)_{|u}) = B(\xi)_{|\tilde{\varphi}(u)}. \tag{1.23}$$

The required result follows then from (1.22), (1.23) and the definition of g_∇.

Finally we need a well-known result of the theory of Lie groups. This result follows from [39, Theorem VIII, chapter IV, p. 105]. Here we give a modified proof.

PROPOSITION 1.9. *Let* M *be a connected and simply connected manifold of dimension* n. *Further let* $X_1,\ldots,X_n$ *be* n *vector fields such that*

i) $X_1,\ldots,X_n$ *are complete;*

ii) $X_1,\ldots,X_n$ *are linearly independent at each point of* M (they determine an absolute parallelism);

iii) $[X_i,X_j] = \sum_k c_{ij}^k X_k$, *where the* c_{ij}^k *are constant.*

Then, for a fixed point $p \in M$, *the manifold* M *has a unique Lie group structure such that* p *is the identity and such that the vector fields* X_i *are all left invariant.* (They constitute a basis for the Lie algebra of the Lie group.)

Proof. Let $\tilde{\nabla}$ denote the linear connection on M defined by

$$\tilde{\nabla}_{X_i}X_j = 0 , \qquad i,j = 1,\ldots,n. \tag{1.24}$$

The curvature of $\tilde{\nabla}$ vanishes and the torsion of $\tilde{\nabla}$ is parallel because

$$\tilde{T}_{X_i}X_j = -[X_i,X_j] = -\sum_k c_{ij}^k X_k .$$

Next let $\mathcal{A}$ denote the Lie algebra generated by $X_1,\ldots,X_n$. $\mathcal{A}$ is a subalgebra of $\mathfrak{X}(M)$. If $X \in \mathcal{A}$, then X is a linear combination of the X_i with constant coefficients and hence X is complete (see [39, Theorem III, p. 95]). Further $\tilde{\nabla}_X X = 0$. This implies that the geodesics of $\tilde{\nabla}$ are the integral curves of fields of $\mathcal{A}$. Hence $\tilde{\nabla}$ is a complete connection.

Let G denote the connected and simply connected Lie group whose Lie algebra g is isomorphic to $\mathcal{A}$. Let $\psi : g \to \mathcal{A}$ denote such an isomorphism and let $(A_1,\ldots,A_n)$ be the basis of g determined by

$$\psi(A_i) = X_i , \qquad i = 1,\ldots,n.$$

Now we consider the (-)-connection $\bar{\nabla}$ of Cartan-Schouten of G (see [26, vol. II, p. 199]). Then

$$\bar{\nabla}_{A_i}A_j = 0 , \quad i,j = 1,\ldots,n$$

and hence $\bar{\nabla}$ has vanishing curvature and parallel torsion. Moreover $\bar{\nabla}$ is complete.

Next we fix a point p of M. Then T_pM is spanned by $X_{|p}$ with $X \in \mathcal{A}$. Hence ψ induces an isomorphism

$$F : g \to T_pM , \; X \mapsto X_{|p}.$$

Identifying g with T_eG we see at once that F preserves the curvature tensor and the torsion tensor of $\bar{\nabla}$ and $\tilde{\nabla}$. Hence there exists an affine transformation $f : G \to M$ such that $f(e) = p$ and $f_{*|e} = F$ (see [26, vol. I, p. 265]). It follows that we can define a Lie group structure on M by putting

$$f(a)f(b) = f(ab) , \qquad a,b \in G. \tag{1.25}$$

Note that (1.25) is equivalent to

$$L_{f(a)} \circ f = f \circ L_a .$$

Next let $X|_p = F(A|_e)$. We shall prove that

$$X = f_*(A). \tag{1.26}$$

Therefore, let $q = f(a)$ and let $\gamma(t)$ be a geodesic joining $a \in G$ to the identity $e \in G$. Denote by $\bar{\tau}_{ea}$ the parallel transport along $\gamma(t)$ with respect to $\bar{\nabla}$ and let $\tilde{\tau}_{pq}$ be the parallel transport along $\tilde{\gamma}(t) = f(\gamma(t))$ with respect to $\tilde{\nabla}$. Put

$$A(t) = \sum_{i=1}^{n} a^i(t) A_i .$$

Then $A(t)$ is parallel along $\gamma(t)$ with respect to $\bar{\nabla}$ if and only if

$$a^i(t) = a^i = \text{constant}$$

since $\bar{\nabla}_{A_i} A_j = 0$. Further, put

$$X(t) = \sum \xi^i(t) X_i .$$

Then $X(t)$ is parallel along $\tilde{\gamma}(t)$ with respect to $\tilde{\nabla}$ if and only if

$$\xi^i(t) = \xi^i = \text{constant} ,$$

since $\tilde{\nabla}_{X_i} X_j = 0$. This shows that

$$\bar{\tau}_{ea}(A|_e) = A|_a , \qquad A \in \mathcal{g} , \tag{1.27}$$

$$\tilde{\tau}_{pq}(X|_p) = X|_q , \qquad X \in \mathcal{A} . \tag{1.28}$$

Since f is an affine isomorphism, it commutes with the parallel transport and hence

$$f_*(A|_a) = f_*(\bar{\tau}_{ea}(A|_e)) = \tilde{\tau}_{pq} f_*(A|_e) =$$

$$= \tilde{\tau}_{pq}(X|_p) = X|_q .$$

This proves (1.26).

Finally this implies that the fields $X \in \mathcal{A}$ are left invariant because

$$
\begin{aligned}
X_{|q} = f_*(A_{|a}) &= f_*((L_a)_* A_{|e}) \\
&= (f \circ L_a)_* A_{|e} \\
&= (L_{f(a)} \circ f)_* A_{|e} \\
&= (L_{f(a)})_* X_{|p}.
\end{aligned}
$$

B. A FIRST PROOF

A connected Riemannian manifold (M,g) is said to be *homogeneous* if the group $\mathfrak{J}(M)$ of isometries acts *transitively* on M. This means : if two points p,q of M are given, then there exists an isometry φ such that $\varphi(p) = q$. Note that in this case the connected component $\mathfrak{J}_0(M)$ of the identity acts transitively on M as well.

We can also say that (M,g) is homogeneous if there exists a connected Lie group G and a C^∞ map

$$G \times M \to M \ , \ (a,p) \mapsto ap = L_a(p)$$

such that for all $a,b \in G$

i) L_a is an isometry of (M,g);

ii) $L_a \circ L_b = L_{ab}$;

iii) for $p,q \in M$, there exists an element $a \in G$ such that $L_a(p) = q$.

In what follows we always suppose that G acts *effectively* on M, i.e. L_a is the identity transformation of M if and only if a is the identity e of G. This is not restrictive since we can always replace G by the quotient group G/N where N is the kernel of the map $a \mapsto L_a$ of G in $\mathfrak{J}(M)$.

If G is a connected Lie group which acts on (M,g) as a

transitive and effective group of isometries, then G can be identified with a Lie subgroup of $\mathcal{J}(M)$ (see for example [8, p. 61]).

Next let $p \in M$ and let $K = K_p = \{a \in G | ap = p\}$ be the *isotropy subgroup* of p. Then M is diffeomorphic to G/K and

$$G \xrightarrow{pr} G/K ,$$

where pr denotes the canonical projection, is a principal fibre bundle over M with structure group K (see [26, vol. I, p. 55]).

Let $\mathfrak{g}$ be the Lie algebra of G and α an element of $\mathfrak{g}$. α^* denotes the vector field on M generated by the one-parameter subgroup

$$q \mapsto (\exp t\alpha)q$$

of G. It is clear that α^* is a Killing vector field and since G acts transitively on M, the function $\mathfrak{g} \to T_pM$, $\alpha \mapsto \alpha^*_{|p}$ is surjective.

More generally, one says that (M,g) is *locally homogeneous* if, for each $p,q \in M$, there exists a neighbourhood $\mathcal{U}$ of p, a neighbourhood $\mathcal{V}$ of q and a local isometry $\varphi : \mathcal{U} \to \mathcal{V}$ such that $\varphi(p) = q$.

In what follows we shall first prove that the existence of a tensor field T satisfying the conditions (AS) implies that (M,g) is locally homogeneous. Then we shall show that (M,g) is homogeneous if some additional topological conditions are fulfilled. Before doing this we need some remarks concerning the equations (AS).

In the first place it is easy to see that these conditions are equivalent to

$$\begin{cases} \text{i)} & \tilde{\nabla} g = 0; \\ \text{ii)} & \tilde{\nabla} R = 0; \\ \text{iii)} & \tilde{\nabla} T = 0 , \end{cases} \tag{1.29}$$

where $\tilde{\nabla}$ is the connection determined by

$$\tilde{\nabla} = \nabla - T. \tag{1.30}$$

Indeed, we have

$$(\tilde{\nabla}_W g)(X,Y) = g(T_W X,Y) + g(X,T_W Y) , \qquad (1.31)$$

$$(\tilde{\nabla}_W T)_X = (\nabla_W T)_X - [T_W,T_X] + T_{T_W X} , \qquad (1.32)$$

$$(\tilde{\nabla}_W R)_{XY} = (\nabla_W R)_{XY} - [T_W,R_{XY}] + R_{T_W XY} + R_{XT_W Y} , \qquad (1.33)$$

$X,Y,W \in \mathfrak{X}(M)$.

Secondly we note that the curvature

$$\tilde{R}_{XY} = \tilde{\nabla}_{[X,Y]} - [\tilde{\nabla}_X,\tilde{\nabla}_Y] , \qquad X,Y \in \mathfrak{X}(M),$$

of $\tilde{\nabla}$ is related to the curvature of ∇ by the formula

$$\tilde{R}_{XY} = R_{XY} + [T_X,T_Y] - T_{T_X Y - T_Y X} , \qquad X,Y \in \mathfrak{X}(M). \qquad (1.34)$$

Hence we have for all $X,Y,W \in \mathfrak{X}(M)$:

$$\begin{aligned}(\tilde{\nabla}_W \tilde{R})_{XY} = {} & (\tilde{\nabla}_W R)_{XY} + [T_X,(\tilde{\nabla}_W T)_Y] - [T_Y,(\tilde{\nabla}_W T)_X] \\ & - (\tilde{\nabla}_W T)_{T_X Y} + (\tilde{\nabla}_W T)_{T_Y X} \\ & - T_{(\tilde{\nabla}_W T)_X Y} + T_{(\tilde{\nabla}_W T)_Y X} .\end{aligned} \qquad (1.35)$$

This implies that the conditions (1.29), and hence the conditions (AS), are equivalent to

$$\begin{cases} \tilde{\nabla} g = 0 , \\ \tilde{\nabla}\tilde{R} = 0 , \\ \tilde{\nabla} T = 0. \end{cases} \qquad (1.36)$$

From this we can conclude that the existence of a solution of the system (AS) is equivalent to the existence of a *metric* connection $\tilde{\nabla}$ which is *invariant under parallelism* (or equivalently, *curvature* and *torsion* of $\tilde{\nabla}$ are *parallel* [36]) and such that the Levi Civita connection ∇ is *rigid with respect to* $\tilde{\nabla}$ (or equivalently, the difference tensor $T = \nabla - \tilde{\nabla}$ is parallel with respect to $\tilde{\nabla}$) (see [26, vol. II, p. 376] and [28]). Note that the *torsion* of $\tilde{\nabla}$ is given by

$$S_X Y = T_Y X - T_X Y , \qquad X, Y \in \mathfrak{X}(M). \tag{1.37}$$

Now we are ready to prove the theorems.

THEOREM 1.10. *Let* (M,g) *be a connected Riemannian manifold such that there exists a tensor field* T *of type* (1,2) *satisfying the conditions* (AS). *Then* (M,g) *is locally homogeneous.*

Proof. Put $\tilde{\nabla} = \nabla - T$ and let p,q be two points of M. Further, let $\gamma(t)$ be a piece-wise differentiable curve which joins p to q. Denote by $\tilde{\tau}_{pq}$ the parallel transport with respect to $\tilde{\nabla}$ along $\gamma(t)$. Since $\tilde{\nabla}$ is metric and invariant under parallelism, $\tilde{\tau}_{pq}$ is an isometry of T_pM on T_qM which preserves the torsion and the curvature of $\tilde{\nabla}$. Hence there exists a neighbourhood $\mathcal{U}$ of p, a neighbourhood $\mathcal{V}$ of q and an affine transformation $\varphi : \mathcal{U} \to \mathcal{V}$ of $\tilde{\nabla}$ such that $\varphi(p) = q$ and $\varphi_{*}|_p = \tilde{\tau}_{pq}$ (see [26, vol. I, p. 262]). The note given after Proposition 1.3 implies the required result.

THEOREM 1.11. *Let* (M,g) *be a connected, simply connected, complete Riemannian manifold satisfying the conditions of Theorem* 1.10. *Then* (M,g) *is homogeneous.*

Proof. Proposition 1.5 implies that $\tilde{\nabla}$ is complete. Then by a standard argument we obtain that the local isometry in Theorem 1.10 can be extended to a global isometry (see [26, vol. I, p. 265]).

In the rest of this section we concentrate on the *converse theorem.*

THEOREM 1.12. *Let* (M,g) *be a homogeneous Riemannian manifold. Then there exists a tensor field* T *of type* (1.2) *satisfying the conditions* (AS).

Proof. We shall give a direct construction of the tensor field T from the group G of isometries of (M,g). Note that the transitive and effective group G need not to be the group $\mathcal{J}(M)$ of all isometries. The construction which we give now is suggested in [27],[28].

Let $\mathfrak{g}$ denote the Lie algebra of G and α an element of $\mathfrak{g}$. We denote by α^* the corresponding fundamental field on M. Recall that α^* is a Killing vector field.

Let $p \in M$ and put

$$\mathfrak{k}_p = \{\alpha \in \mathfrak{g} \mid \alpha^*_{|p} = 0\}. \tag{1.38}$$

$\mathfrak{k}_p$ is a Lie subalgebra of $\mathfrak{g}$ since (see [26, vol. II, p. 469])

$$[\alpha^*, \beta^*] = -[\alpha,\beta]^*. \tag{1.39}$$

We shall see that $\mathfrak{k}_p$ is the Lie algebra of the isotropy subgroup K_p of p. At this moment it is sufficient to note that, if Ad is the adjoint representation of G on $\mathfrak{g}$, we have

$$(a \exp t\alpha)p = (a \exp t\alpha\, a^{-1})ap$$

and hence

$$(\mathrm{Ad}(a)\alpha)^*_{|ap} = (L_a)_{*|p}(\alpha^*_{|p}) \,. \tag{1.40}$$

Now let $V = T_pM$ be the tangent vector space of M at p. The *Killing form* B of $\mathfrak{so}(V)$ is given by

$$B(\varphi,\psi) = \mathrm{tr}(\varphi \circ \psi) \tag{1.41}$$

where φ and ψ are skew-symmetric endomorphisms of V. B is a symmetric bilinear form which is negative definite. Since α^* is a Killing vector field, A_{α^*} is skew-symmetric (see section A). Hence we can put

$$\phi_p(\alpha,\beta) = -B(A_{\alpha^*|p}, A_{\beta^*|p}) \,, \qquad \alpha,\beta \in \mathfrak{g}. \tag{1.42}$$

ϕ_p is a symmetric bilinear form on $\mathfrak{g}$ which depends on p. Moreover we have

LEMMA 1.13. *ϕ_p is positive definite on k_p.*

Proof. Let $\beta \in k_p$ with $\phi_p(\beta,\beta) = 0$. Then $A_{\beta^*|p} = 0$. Since $\beta^*_{|p} = 0$ it follows from Proposition 1.2 that β^* is identically zero. Finally, since G acts effectively on M, we have $\beta = 0$.

LEMMA 1.14. *For all $\alpha,\beta \in g$ we have*

$$\phi_p(\mathrm{Ad}(a)\alpha,\mathrm{Ad}(a)\beta) = \phi_{a^{-1}p}(\alpha,\beta) , \qquad p \in M,\ a \in G,$$

and ϕ_p is $\mathrm{Ad}(k_p)$-invariant.

Proof. (1.41) and (1.42) imply :

$$\phi_p(\mathrm{Ad}(a)\alpha,\mathrm{Ad}(a)\beta) = -B(A_{\mathrm{Ad}(a)\alpha^*|p}, A_{\mathrm{Ad}(a)\beta^*|p})$$

$$= -\sum_m g((A_{\mathrm{Ad}(a)\alpha^*|p} \circ A_{\mathrm{Ad}(a)\beta^*|p})e_m, e_m) ,$$

where $(e_m, m = 1,\ldots,n)$ is an orthonormal basis of $V = T_pM$. Since A_ξ is skew-symmetric for a Killing vector field ξ we have

$$\phi_p(\mathrm{Ad}(a)\alpha,\mathrm{Ad}(a)\beta) = \sum_m g(\nabla_{e_m}(L_a)_*(\alpha^*)_{|p}, \nabla_{e_m}(L_a)_*(\beta^*)_{|p}). \qquad (1.43)$$

Further, L_a is an isometry of (M,g) and so (1.2) implies

$$\nabla_{e_m}(L_a)_*(\alpha^*)_{|p} = (L_a)_*(\nabla_{(L_a)_*^{-1}e_m}\alpha^*)_{|a^{-1}p} . \qquad (1.44)$$

Hence we have

$$\phi_p(\mathrm{Ad}(a)\alpha,\mathrm{Ad}(a)\beta) = \sum_m g(\nabla_{(L_a)_*^{-1}e_m}\alpha^*_{|a^{-1}p}, \nabla_{(L_a)_*^{-1}e_m}\beta^*_{|a^{-1}p}).$$

Finally, since the vectors $(L_a)_*^{-1}e_m$, $m = 1,\ldots,n$, form an orthonormal basis of $T_{a^{-1}p}M$, the result now follows at once.

Next we put

$$m_p = \{\alpha \in g \,|\, \phi_p(\alpha,\beta) = 0 \text{ for all } \beta \in k_p\}. \tag{1.45}$$

m_p is a vector subspace of g. Moreover we have

LEMMA 1.15. (i) $g = k_p \oplus m_p$ *(direct sum)*;

(ii) *the linear map* $m_p \to T_pM$ $,\alpha \mapsto \alpha_p^*$ *is an isomorphism;*

(iii) k_p *is the Lie algebra of the isotropy subgroup of* G *at* p;

(iv) $Ad(a)m_p \subseteq m_p$ *for all* $a \in G$;

(v) $Ad(k_p)m_p \subseteq m_p$.

Proof. (i) Let $\alpha \in k_p \cap m_p$. Then $\phi_p(\alpha,\alpha) = 0$ and hence $\alpha = 0$ (lemma 1.13). Moreover, if $(u_1,\ldots,u_r)$ is a basis of k_p such that $\phi_p(u_m,u_\ell) = \delta_{m\ell}$, then we have

$$\alpha - \sum_{\ell=1}^{r} \phi_p(u_\ell,\alpha)u_\ell \in m_p$$

for each $\alpha \in g$.

(ii) Let $X \in T_pM$. Then there exists an element γ of g such that $\gamma^*|_p = X$ since G acts transitively on M. But $\gamma = \alpha + \beta$ where $\alpha \in m_p$ and $\beta \in k_p$. Hence $\gamma_p^* = \alpha^*|_p = X$ and α^* is the unique element with this property.

(iii) This follows immediately from (i) and (ii) because of the dimensions and the fact that the Lie algebra of the isotropy subgroup of p is contained in k_p.

(iv) Let $a \in G$ and $\alpha \in m_p$. Then Lemma 1.14 implies

$$\phi_{ap}(Ad(a)\alpha,\beta) = \phi_p(\alpha,Ad(a^{-1})\beta)$$

for all $\beta \in k_{ap}$. Further $Ad(a^{-1})\beta \in k_p$ and hence

$$\phi_{ap}(\mathrm{Ad}(a)\alpha,\beta) = 0.$$

This gives the required result.

(v) This follows at once from (iv).

<u>REMARK</u>. Lemma 1.15 (v) shows that all homogeneous Riemannian manifolds are *reductive* (see [27] and [26, vol. II, p. 190]. See also chapter 6.)

Now we finish the proof of the theorem. Let $\tilde{\nabla}$ be the *canonical connection* of the reductive homogeneous space $M = G/K$ with respect to the decomposition

$$\mathfrak{g} = \mathfrak{m}_p \oplus \mathfrak{k}_p$$

of the Lie algebra $\mathfrak{g}$ of G (see [26, vol. II, p. 192]). $\tilde{\nabla}$ is a metric connection which is invariant under the action of G. This connection is uniquely determined by

$$(\tilde{\nabla}_{\alpha^*}\beta^*)_{|p} = [\alpha^*,\beta^*]_{|p} = - [\alpha,\beta]^*_{|p} ,$$

for $\alpha,\beta \in \mathfrak{g}$.

Further, let $T = \nabla - \tilde{\nabla}$ where ∇ is the Levi Civita connection. Since G acts by isometries, ∇ is also G-invariant. Hence T is G-invariant. Then T is parallel with respect to $\tilde{\nabla}$ (see [26, vol. II, p. 193]). For the same reasons the curvature R is parallel. So T satisfies the conditions (AS) and we obtain the required result.

Note that T is uniquely determined by its value at p because T is G-invariant. This value is given by

$$(T_{\alpha^*}\beta^*)_{|p} = \nabla_{\beta^*}\alpha^*_{|p} = - A_{\alpha^*}(\beta^*)_{|p} \tag{1.46}$$

for $\alpha,\beta \in \mathfrak{g}$. This means that if $X,Y \in T_pM$ and if α is the unique element of $\mathfrak{m}_p$ such that $\alpha^*_{|p} = X$, we have

$$(T_p)_X Y = \nabla_Y \alpha^*_{|p} = - A_{\alpha^*|p}(Y). \tag{1.47}$$

C. THE PROOF OF AMBROSE AND SINGER

In this final section we reproduce, with slight modifications, the proof given by Ambrose and Singer. In contrast to the procedure used in section B, we shall start with the construction of the tensor T when a transitive and effective group of isometries of (M,g) is given. In that way we find a different proof of Theorem 1.12.

Let G be a connected group of isometries acting transitively and effectively on the connected Riemannian manifold (M,g) and let $\mathcal{O}(M)$ be the bundle of orthonormal frames with the metric g_∇ as constructed in section A. Then Proposition 1.8 implies that G also acts as a group of isometries on $(\mathcal{O}(M), g_\nabla)$. This action is given by

$$L_a u = au = a(p;u_1,\ldots,u_n) = (ap;(L_a)_* u_1,\ldots,(L_a)_* u_n) , \tag{1.48}$$

$a \in G$. This action is a *free action* because if $au = u$ for some $a \in G$, then $L_a(p) = p$ and $(L_a)_*|_p$ is the identity transformation of T_pM. Then Proposition 1.1 implies that L_a is the identity on M. But G is effective and hence $a = e$.

For each $u \in \mathcal{O}(M)$ define J_u by

$$J_u : G \to \mathcal{O}(M) , \; a \mapsto au. \tag{1.49}$$

J_u is a C^∞ function and since G acts freely, J_u is *injective* and also *non-singular*. Hence J_u is a *one-to-one immersion* and J_uG is a *submanifold* of $\mathcal{O}(M)$.

Let $\mathfrak{g}$ be the Lie algebra of G identified with T_eG and put

$$G_u = (J_u)_* \mathfrak{g} \subset T_u\mathcal{O}(M). \tag{1.50}$$

Let V_u denote the vertical space at u and $(G_u \cap V_u)^\perp$ the orthogonal complement of $G_u \cap V_u$ in $T_u\mathcal{O}(M)$ with respect to g_∇. Put

$$Q_u = G_u \cap (G_u \cap V_u)^\perp.$$

Then we have

LEMMA 1.16. (i) $Q_u \oplus V_u = T_u\mathcal{O}(M)$ (*direct sum*);

(ii) $(R_a)_* Q_u = Q_{ua}$;

(iii) Q_u *depends differentiably on* u.

<u>Proof</u>. (i) We prove that π_* is an isomorphism of Q_u onto T_pM where $p = \pi(u)$. Let $X \in T_pM$. Since G is a transitive group, there exists an element $\alpha \in g$ such that $\alpha^*|_p = X$. Hence X is tangent to the curve $(\exp t\alpha)p$ at $t = 0$. (1.48) and (1.49) imply

$$(\exp t\alpha)p = \pi((\exp t\alpha)u) = (\pi \circ J_u)(\exp t\alpha)$$

and hence

$$X = \pi_*(J_u)_*(\alpha).$$

This shows that π_* is surjective. On the other hand, if $\pi_*(J_u)_*(\alpha) = 0$ and $(J_u)_*(\alpha) \in Q_u$, we have

$$(J_u)_*(\alpha) \in (G_u \cap V_u)^\perp \cap (G_u \cap V_u)$$

and hence $\alpha = 0$. So π_* is injective.

(ii) As proved in Proposition 1.7, $R_b : \mathcal{O}(M) \to \mathcal{O}(M)$ is an isometry of $(\mathcal{O}(M), g_\nabla)$ for all $b \in O(n)$. Hence, since

$$J_{ub} = R_b \circ J_u ,$$

we have

$$G_{ub} = (R_b)_* G_u .$$

But we also have

$$V_{ub} = (R_b)_* V_u$$

and hence

$$Q_{ub} = (R_b)_* Q_u$$

because $(R_b)_*$ preserves the orthogonal complement.

(iii) Let $\alpha \in \mathfrak{g}$. $(J_u)_*\alpha$ is the value at u of the vector field induced on $\mathcal{O}(M)$ by the action of G. Hence, if α is fixed, $(J_u)_*\alpha$ depends differentiably on u. This shows that $u \mapsto G_u$ is a C^∞ distribution. Since $u \mapsto V_u$ is also C^∞ we find that $G_u \cap V_u$, its orthogonal complement and hence Q_u are also C^∞ distributions.

This lemma shows that $u \mapsto Q_u$ is an *infinitesimal connection* on $\mathcal{O}(M)$. Moreover, an important fact will be that this connection is *invariant* under the action of G. Indeed we have

LEMMA 1.17. *For any* $a \in G$ *and any* $u \in \mathcal{O}(M)$ *we have*

$$(L_a)_*Q_u = Q_{au}.$$

Proof. First we note that

$$L_a \circ J_u = J_u \circ L_a \tag{1.51}$$

for $a \in G$ and $u \in \mathcal{O}(M)$. So we have for $\alpha \in \mathfrak{g}$:

$$\begin{aligned}(L_a)_*|_u(J_u)_*|_e(\alpha|_e) &= (L_a \circ J_u)_*|_e(\alpha|_e) = \\ &= (J_u \circ L_a)_*|_e(\alpha|_e) = (J_u)_*|_a(L_a)_*(\alpha|_e) = \\ &= (J_u)_*|_a(\alpha|_a)\end{aligned}$$

since α is left invariant. But $(J_u)_*|_a(\alpha|_a)$ is tangent at $au = J_u(a)$ to the curve

$$a\,\exp t\alpha\, u = (\mathrm{ad}(a)\exp t\alpha)au = (\exp t\mathrm{Ad}(a)\alpha)au.$$

Hence

$$(J_u)_*|_a(\alpha|_a) = (J_{au})_*(\mathrm{Ad}(a)\alpha). \tag{1.52}$$

This shows that

$$(L_a)_* G_u = G_{au}. \tag{1.53}$$

On the other hand we have

$$\pi \circ L_a = L_a \circ \pi \tag{1.54}$$

and so

$$(L_a)_* V_u = V_{au}.$$

Finally, Proposition 1.8 implies that L_a is an isometry of $(\mathcal{O}(M),g_\nabla)$ and so the required result follows at once.

With this infinitesimal connection $u \mapsto Q_u$ we can associate in a canonical way a *metric linear connection* $\tilde{\nabla}$ which is G-*invariant* (see [26]). The difference tensor $T = \nabla - \tilde{\nabla}$ is also G-invariant. Hence, to prove that T satisfies the conditions (AS) it suffices to show that at a point $p \in M$ we have

$$T_X Y = \nabla_Y \alpha^*|_p \tag{1.55}$$

where $X,Y \in T_pM$ and α is the unique element of m_p such that $\alpha^*|_p = X$. The required result then follows from the results obtained in section B.

First we note that

$$Q_u = \{(J_u)_*(\alpha) \,|\, \alpha \in m_p\}. \tag{1.56}$$

Indeed, $(J_u)_*$ is injective and we have already seen that $\dim Q_u = \dim m_p = \dim M = n$. On the other hand we know that $(J_u)_*\alpha$ is tangent at $t = 0$ to the curve

$$\begin{aligned} \gamma(t) &= (\exp t\alpha)u \\ &= ((\exp t\alpha)p; (L_{\exp t\alpha})_* u_1, \ldots, (L_{\exp t\alpha})_* u_n). \end{aligned} \tag{1.57}$$

Let $\mathcal{U}$ be a neighbourhood of p and $(E_1,\ldots,E_n)$ a local section of $\mathcal{O}(M)$ on $\mathcal{U}$. We identify $\pi^{-1}(\mathcal{U})$ with $\mathcal{U} \times O(n)$. Then $u = (p,a)$, where a is the element (a_h^k) of $O(n)$ such that

$$u_h = \sum_k a_h^k E_k|_p. \qquad (1.58)$$

Similarly we have

$$\gamma(t) = ((\exp t\alpha)p, a(t))$$

where a(t) is the curve in O(n) determined by

$$a_h^k(t) = g_{(\exp t\alpha)p}((L_{\exp t\alpha})_* u_h, E_k)$$

$$= \sum_m a_h^m g_{(\exp t\alpha)(p)}((L_{\exp t\alpha})_* E_m, E_k)$$

$$= \sum_m a_h^m g_p(E_m, (L_{\exp t\alpha})_*^{-1} E_k)$$

because $L_{\exp t\alpha}$ is an isometry.

From the definition of the Lie derivative we obtain (see [26, vol. I]) :

$$\dot{a}_h^k(0) = \sum_m a_h^m g_p(E_m, [\alpha^*, E_k]|_p). \qquad (1.59)$$

Now put

$$A_h^k = g(E_h, [\alpha^*, E_k]|_p). \qquad (1.60)$$

Then $A = (A_h^k) \in so(n)$. Hence we can write

$$(J_u)_* \alpha = \alpha^*|_p + Aa. \qquad (1.61)$$

It follows from (1.12), (1.13) and (1.14) that

$$\tilde{\omega}(\alpha^*|_p) = -A. \qquad (1.62)$$

So, if $X = \alpha^*|_p$, we have

$$\tilde{\omega}_i^j(X) = g_p(\tilde{\nabla}_X E_i, E_j) = -A_i^j =$$

$$= -g_p(E_i, [\alpha^*, E_j]|_p) = g_p(E_j, [\alpha^*, E_i]|_p)$$

and hence

$$\tilde{\nabla}_X E_i|_p = [\alpha^*, E_i]|_p . \tag{1.63}$$

Finally, using (1.63), we obtain

$$(T_X E_i)|_p = \nabla_X E_i|_p - \tilde{\nabla}_X E_i|_p$$

$$= (\nabla_{\alpha^*} E_i - \tilde{\nabla}_{\alpha^*} E_i)|_p = \nabla_{E_i} \alpha^*|_p$$

and this proves the required result.

We finish this section with the proof of the *converse theorem*. Let M be a connected, simply connected and complete Riemannian manifold and suppose there exists a tensor field T of type (1,2) which satisfies the conditions (AS). Further let

$$\tilde{\nabla} = \nabla - T.$$

Then $\tilde{\nabla}$ is a metric linear connection. The result then follows from the following :

THEOREM 1.18. *Let* $\mathcal{O}(M)$ *be the bundle of orthonormal frames on* M *and* u *a point of* $\mathcal{O}(M)$. *If* $\tilde{\mathcal{T}}_u$ *denotes the holonomy bundle of* $\tilde{\nabla}$ *through* u, *then* $\tilde{\mathcal{T}}_u = G$ *is connected, has a Lie group structure and* $\tilde{\mathcal{T}}_u$ *acts on* M *as a transitive and effective group of isometries.*

Proof. Recall that $\tilde{\mathcal{T}}_u$ is the set of all $v \in \mathcal{O}(M)$ which can be joined to u by a piece-wise differentiable horizontal curve and hence is connected. Moreover, $\tilde{\mathcal{T}}_u$ is a principal subbundle of $\mathcal{O}(M)$ whose structure group is the *holonomy group* $\tilde{\psi}_u$ of $\tilde{\nabla}$. This group can be identified with a subgroup of O(n) : it is the group of isometries of T_pM, $p = \pi(u)$, obtained by parallel transport with respect to $\tilde{\nabla}$ along the loops through p.

Let $(A_1,\ldots,A_r)$, $r = \dim \tilde{\psi}_u$, be a basis for the Lie algebra of $\tilde{\psi}_u$ which is identified with a subalgebra of $\mathfrak{so}(n)$. Recall that the corresponding vertical fields $A_1^*,\ldots,A_r^*$ are complete. Further, let $B_1,\ldots,B_n$ be the standard horizontal vector fields with respect to $\tilde{\nabla}$ and corresponding to a natural basis of $\mathbb{R}^n$. These fields are also complete since $\widetilde{\nabla}$ is complete (Proposition 1.5 and [26, vol. I, p. 140]).

The restrictions of $B_1,\ldots,B_n,A_1^*,\ldots,A_r^*$ determine an absolute parallelism on $\widetilde{\mathcal{T}}_u$ and are still complete. Moreover, we have (see [26, vol. I, p. 42 and p. 120]) :

$$[A_i^*,A_j^*] = [A_i,A_j]^* , \tag{1.64}$$

$$[A_i^*,B(\xi)] = B(A(\xi)) , \quad \xi \in \mathbb{R}^n. \tag{1.65}$$

Further, let $\widetilde{R}$ denote the curvature and $\widetilde{S}$ the torsion of $\widetilde{\nabla}$. Since $\widetilde{\nabla}$ is invariant under parallelism, i.e. $\widetilde{\nabla}\widetilde{R} = \widetilde{\nabla}\widetilde{S} = 0$, we have on $\widetilde{\mathcal{T}}_u$ ([26, vol. I, p. 137]) :

$$\begin{aligned}[B(\xi),B(\eta)] &= -\ 2B(\Theta(B(\xi),B(\eta)) \\ &\quad -\ 2(\Omega(B(\xi),B(\eta)))^* , \qquad \xi,\eta \in \mathbb{R}^n ,\end{aligned} \tag{1.66}$$

where Θ is the *torsion form* and Ω the *curvature form* of the connection $\widetilde{\nabla}$. Note that $\Theta(B(\xi),B(\eta))$ is a constant function (with values in $\mathbb{R}^n$) on $\widetilde{\mathcal{T}}_u$ and that $\Omega(B(\xi),B(\eta))$ is a constant function (with values in the Lie algebra $\widetilde{\psi}_u$) on $\widetilde{\mathcal{T}}_u$.

It follows (corollary 5.6 of [26, vol. I, p. 138]) that, on $\widetilde{\mathcal{T}}_u$, the brackets (1.64), (1.65) and (1.66) are *constant linear combinations* of the restrictions of $B_1,\ldots,B_n,A_1^*,\ldots,A_r^*$. Hence these fields generate a Lie subalgebra of $\mathfrak{X}(\widetilde{\mathcal{T}}_u)$ of dimension n + r. Let $\mathfrak{g}$ denote these Lie algebra.

Next let $\bar{G}$ be the universal covering of $\widetilde{\mathcal{T}}_u$ and ρ the projection of $\bar{G}$ on $\widetilde{\mathcal{T}}_u$. Further, let $\bar{B}_i,\bar{A}_\alpha^*$ be the vector fields on $\bar{G}$ which are uniquely determined by

$$\begin{aligned} \rho_*(\bar{B}_i) &= B_i , \qquad i = 1,\ldots,n ; \\ \rho_*(\bar{A}_\alpha^*) &= A_\alpha^* , \qquad \alpha = 1,\ldots,r . \end{aligned} \tag{1.67}$$

Now $\bar{G}$ satisfies the hypotheses of Proposition 1.9. Hence, if $\bar{e}$ is a fixed point of $\bar{G}$ (we always suppose $\bar{e} \in \rho^{-1}(u)$), then there exists a unique Lie group structure on $\bar{G}$ such that $\bar{e}$ is the identity and such that

$\bar{B}_1,\ldots,\bar{B}_n,\bar{A}_1^*,\ldots,\bar{A}_r^*$ are left invariant vector fields. Note that $\bar{A}_1^*,\ldots,\bar{A}_r^*$ generate a subalgebra $\mathcal{g}_0$ of $\mathcal{g}$ (see (1.64)). Let $\bar{G}_0$ be the connected subgroup of $\bar{G}$ with Lie algebra $\mathcal{g}_0$. We first prove

LEMMA 1.19. *M and* $\bar{G}/\bar{G}_0$ *are diffeomorphic.*

Proof. Let

$$\pi_1 = \pi \circ \rho. \tag{1.68}$$

Then $\pi_1 : \bar{G} \to M$ is a fibre bundle with projection map π_1. Further, it follows from the bundle homotopy sequence (see [46, p. 377]) and from the fact that M is simply connected, that the fibres are connected. Since π_1 is continuous, they are also closed. On the other hand we have

$$(\pi_1)_*(\bar{A}_\alpha^*) = 0 , \qquad \alpha = 1,\ldots,r , \tag{1.69}$$

and hence the fibres are tangent to $\mathcal{g}_0$. As a consequence the fibres are maximal integral submanifolds of the involutive distribution determined by $\mathcal{g}_0$. So the fibres are the classes $\bar{a}\bar{G}_0$. The projection map π_1 induces a C^∞ function

$$\pi_2 : \bar{G}/\bar{G}_0 \to M$$

which is a diffeomorphism since $(\pi_2)_*$ is an isomorphism at each point and π_2 is a 1 - 1 function.

Next let $q \in M$. We can always write

$$q = \bar{b}\bar{G}_0$$

where $\bar{b} \in \bar{G}$. If $v = \rho(\bar{b})$ is a point of $\tilde{\mathcal{T}}_u$, we have

$$v = (\bar{b}\bar{G}_0; v_1,\ldots,v_n) \tag{1.70}$$

where the vectors v_i are given by

$$v_i = (\pi_*)|_v(B_i|_v) = (\pi_1)_*|_{\bar{b}}(\bar{B}_i|_{\bar{b}}). \tag{1.71}$$

Hence, all the points of $\tilde{\mathcal{T}}_u$ are of the form

$$\rho(\bar{b}) = (\bar{b}\bar{G}_0; (\pi_1)_{*|\bar{b}}(\bar{B}_1|\bar{b}), \ldots, (\pi_1)_{*|\bar{b}}(\bar{B}_n|\bar{b})) \tag{1.72}$$

where $\bar{b} \in \bar{G}$. Now we have

<u>LEMMA 1.20.</u> *The left translations*

$$L_{\bar{a}} : M \to M \,, \quad \bar{b}\bar{G}_0 \mapsto \bar{a}\bar{b}\bar{G}_0$$

are isometries of M.

<u>Proof</u>. Let $\mathcal{L}_{\bar{a}}$ denote the left translation of $\bar{G}$ corresponding to $\bar{a}$. We have

$$L_{\bar{a}} \circ \pi_1 = \pi_1 \circ \mathcal{L}_{\bar{a}} \,. \tag{1.73}$$

Hence, if

$$q = \pi_1(\bar{b}) = \bar{b}\bar{G}_0 = \pi(\rho(\bar{b})) \quad ,$$

we have

$$\begin{aligned}(L_{\bar{a}})_{*|q}(\pi_1)_{*|\bar{b}}(\bar{B}_i|\bar{b}) &= (\pi_1 \circ \mathcal{L}_{\bar{a}})_{*|\bar{b}}(\bar{B}_i|\bar{b}) = \\ (\pi_1)_{*|\bar{a}\bar{b}}((\mathcal{L}_{\bar{a}})_{*|\bar{b}}(\bar{B}_i|\bar{b})) &= (\pi_1)_{*|\bar{a}\bar{b}}(\bar{B}_i|\bar{a}\bar{b}) \,,\end{aligned} \tag{1.74}$$

$i = 1,\ldots,n$, since the fields $\bar{B}_i$ are left invariant. The vectors $(\pi_1)_{*|\bar{b}}(\bar{B}_i|\bar{b})$ form an orthonormal basis of T_qM (see (1.71) and for the same reason the vectors $(\pi_1)_{*|\bar{a}\bar{b}}(\bar{B}_i|\bar{a}\bar{b})$ form an orthonormal basis of $T_{L_{\bar{a}}q}M$. This and (1.74) proves that $L_{\bar{a}}$ is an isometry.

To finish the proof of the theorem we need a further result.

<u>LEMMA 1.21.</u> *Let* $\bar{a} \in \bar{G}$ *and denote by* $\tilde{L}_{\bar{a}}$ *the diffeomorphism of* $\mathcal{O}(M)$ *induced by* $L_{\bar{a}}$ *(see Proposition* 1.8*). Then*

$$\tilde{L}_{\bar{a}}(\tilde{\mathcal{T}}_u) \subset \tilde{\mathcal{T}}_u \,.$$

Proof. Let $\rho(\bar{b}) \in \tilde{\mathcal{T}}_u$. Then from (1.72) we have

$$\rho(\bar{b}) = (\bar{b}\bar{G}_0; (\pi_1)_{*|\bar{b}}(\bar{B}_1|_{\bar{b}}), \ldots, (\pi_1)_{*|\bar{b}}(\bar{B}_n|_{\bar{b}})) .$$

From the definition of $\tilde{L}_{\bar{a}}$ we get :

$$\tilde{L}_{\bar{a}}(\rho(\bar{b})) = (L_{\bar{a}}(\bar{b}\bar{G}_0); (L_{\bar{a}})_*(\pi_1)_{*|\bar{b}}(\bar{B}_1|_{\bar{b}}), \ldots, (L_{\bar{a}})_*(\pi_1)_{*|\bar{b}}(\bar{B}_n|_{\bar{b}}))$$

and then, (1.74) implies

$$\tilde{L}_{\bar{a}}(\rho(\bar{b})) = (\bar{a}\bar{b}\bar{G}_0; (\pi_1)_{*|\bar{a}\bar{b}}(\bar{B}_1|_{\bar{a}\bar{b}}), \ldots, (\pi_1)_{*|\bar{a}\bar{b}}(\bar{B}_1|_{\bar{a}\bar{b}})).$$

Hence $\tilde{L}_{\bar{a}}(\rho(\bar{b})) = \rho(\bar{a}\bar{b}) \in \tilde{\mathcal{T}}_u$.

We conclude from this lemma that $\bar{G}$ acts on $\tilde{\mathcal{T}}_u$. Moreover it acts *transitively* because

$$\tilde{L}_{\bar{b}_1\bar{b}_2^{-1}}(\rho(\bar{b}_2)) = \tilde{L}_{\bar{b}_1\bar{b}_2^{-1}}(\bar{b}_2\bar{G}_0; (\pi_1)_{*|\bar{b}_2}(\bar{B}_1|_{\bar{b}_2}), \ldots, (\pi_1)_{*|\bar{b}_2}(\bar{B}_n|_{\bar{b}_2}))$$

$$= (\bar{b}_1\bar{G}_0; (\pi_1)_{*|\bar{b}_1}(\bar{B}_1|_{\bar{b}_1}), \ldots, (\pi_1)_{*|\bar{b}_1}(\bar{B}_n|_{\bar{b}_1}))$$

$$= \rho(\bar{b}_1).$$

Next recall that the identity $\bar{e}$ of $\bar{G}$ has been chosen in $\rho^{-1}(u)$. We now prove that the isotropy group $\bar{K}$ of $u \in \tilde{\mathcal{T}}_u$ is the kernel of the map $\bar{a} \mapsto L_{\bar{a}}$ of $\bar{G}$ into $\mathcal{J}(M)$. Indeed, if $\bar{a} \in \bar{K}$, then $\tilde{L}_{\bar{a}}(u) = u$. Using the definition of $L_{\bar{a}}$ and (1.72) we get from this :

$$\begin{cases} L_{\bar{a}}(\pi(u)) = \pi(u) , \\ (L_{\bar{a}})_*((\pi_1)_{*|\bar{e}}(\bar{B}_i|_{\bar{e}})) = (\pi_1)_{*|\bar{e}}(\bar{B}_i|_{\bar{e}}) , \qquad i = 1,\ldots,n. \end{cases} \tag{1.75}$$

Hence $L_{\bar{a}}$ fixes $p = \pi(u) \in M$ and $(L_{\bar{a}})_{*|p}$ is the identity transformation of T_pM. Proposition 1.1 implies that $L_{\bar{a}}$ is the identity on M and so $\bar{a}$ is an element of the kernel of $\bar{a} \mapsto L_{\bar{a}}$. The converse is obvious because if $L_{\bar{a}}$ is the identity on M, then $\tilde{L}_{\bar{a}}$ is the identity on $\tilde{\mathcal{T}}_u$. Hence $\bar{K}$ is a *normal* subgroup of $\bar{G}$ which acts trivially on M. This implies that $\tilde{\mathcal{T}}_u = \bar{G}/\bar{K}$ has a Lie group structure and its acts transitively and effectively on M.

Note that $\bar{K}$ is *discrete* because $\bar{G}$ is a covering of $\tilde{\mathcal{T}}_u$. Hence the action of $\bar{G}$ on M is *almost effective* (see [26, vol. II, p. 187]).

REMARK. The proof of Theorem 1.18 shows that the Lie algebra g of $G = \tilde{\mathcal{T}}_u$ is a direct sum

$$g = m \oplus k \tag{1.76}$$

where $k = g_0$ is generated by the restrictions of $A_1^*,\dots,A_r^*$ on G and m is the vector subspace of g spanned by the restrictions of $B_1,\dots,B_n$.

For what follows in these notes it will be important to have a simpler construction for this Lie algebra. Therefore recall that $\Theta(B(\xi),B(\eta))$ is constant on G, where Θ is the torsion form of $\tilde{\nabla}$. Hence it suffices to compute it in e = u of G. To simplify the notations we identify T_pM, $p = \pi(e)$, with $\mathbb{R}^n$ using the isomorphism

$$u : \mathbb{R}^n \to T_pM , \quad (\xi^1,\dots,\xi^n) \mapsto \sum_{i=1}^{n} \xi^i u_i .$$

Then one has (see [26, vol. I, p. 132]) :

$$2\Theta(B(\xi),B(\eta)) = 2\Theta(B(\xi),B(\eta))_{|u} = (\tilde{S}_p)_\xi\eta \tag{1.77}$$

where $\tilde{S}$ is the torsion tensor of $\tilde{\nabla}$.

Similarly we have

$$2\Omega(B(\xi),B(\eta)) = -(\tilde{R}_p)_{\xi\eta} , \tag{1.78}$$

where $\tilde{R}$ is the curvature tensor of $\tilde{\nabla}$ (see [26, vol. I, p. 133]).

Now we identify m with $\mathbb{R}^n$ by the map $B(\xi) \mapsto \xi$ and k with the *holonomy algebra* (the Lie algebra of $\tilde{\psi}_u$) by $A^* \mapsto A$. Recall that k is generated by the skew-symmetric endomorphisms of $T_pM \to \mathbb{R}^n$ of type $(\tilde{R}_p)_{\xi\eta}$ (see [26, vol. I, p. 151]). Hence the Lie algebra g can be identified via the isomorphism

$$B(\xi) + A^* \mapsto \xi + A$$

with the direct sum of $\mathbb{R}^n$ and the subalgebra of $\mathfrak{so}(n)$ generated by the $(\tilde{R}_p)_{\xi\eta}$, equipped with the following brackets :

$$\begin{cases} [A,A'] = AA' - A'A \,, \\ [A,\xi] = A(\xi) \,, \\ [\xi,\eta] = (T_p)_\xi\eta - (T_p)_\eta\xi + (\tilde{R}_p)_{\xi\eta} \end{cases} \tag{1.79}$$

where $A,A' \in \mathfrak{so}(n)$ and $\xi,\eta \in \mathbb{R}^n$. This follows from (1.37), (1.64), (1.65), (1.66), (1.77) and (1.78).

Note that (1.79) implies also that M is a *reductive homogeneous space* and that $\tilde{\nabla} = \nabla - T$ is the *canonical connection* with respect to the reductive decomposition (1.76) (see chapter 6).

In the rest of these notes we shall always identify $\mathfrak{g}$ with this Lie algebra and $G = \tilde{\mathcal{T}}_u$ with the connected Lie subgroup of $\mathfrak{J}(M)$ whose Lie algebra is isomorphic to $\mathfrak{g}$.

Using the remarks above, we are now able to give an answer to the problem stated in the introduction : Which groups G' can be obtained from the solutions of the Ambrose-Singer equations ? Therefore we need the notion of group of transvections.

<u>DEFINITION 1.22.</u> Let $(M,\bar{\nabla})$ be a connected manifold with affine connection $\bar{\nabla}$. The group of all affine transformations of M preserving each holonomy bundle $\bar{\mathcal{T}}_u$ is called the *group of transvections* of $(M,\bar{\nabla})$.

Note that if an affine transformation preserves $\bar{\mathcal{T}}_{u_0}$ for some u_0, then it also preserves the holonomy bundle $\bar{\mathcal{T}}_u$ for all u.

More geometrically, an affine transformation φ of $(M,\bar{\nabla})$ belongs to the group of transvections if and only if the following holds : for every point $m \in M$ there is a piece-wise differentiable curve γ joining m to $\varphi(m)$ such that the map $\varphi_{*|m} : T_mM \to T_{\varphi(m)}M$ coincides with the parallel transport along γ.

Using Theorem I.25 of [30] we see that the group G' obtained from the solution T and the corresponding reductive decomposition of $\mathfrak{g}$ is the group of transvections of $(M,\tilde{\nabla})$ where $\tilde{\nabla}$ is the corresponding canonical metric connection.

We refer to the end of chapter 2 for further information about this problem. See also [30, p. 41].

2. HOMOGENEOUS RIEMANNIAN STRUCTURES

In this chapter we consider the notion of "homogeneous Riemannian structure" and prove a characterization of "isomorphic" homogeneous Riemannian structures. (M,g) always denotes a Riemannian manifold of class C^∞.

DEFINITION 2.1. A *homogeneous (Riemannian) structure* on (M,g) is a tensor field T of type $(1,2)$ which is a solution of the system (AS).

First we recall that the existence of a homogeneous structure T on (M,g) does not imply that this manifold is homogeneous even if one supposes the manifold to be connected and complete. In that case one can only say that the manifold is *locally homogeneous* (see chapter 1, section B). This is easily seen if one considers the universal Riemannian covering space $(\widetilde{M},\widetilde{g})$ of (M,g). It is clear that the homogeneous structure T on (M,g) induces a homogeneous structure $\widetilde{T}$ on $(\widetilde{M},\widetilde{g})$. Hence if M is complete and connected, then $\widetilde{M}$ is connected, complete and simply connected and so $(\widetilde{M},\widetilde{g})$ is homogeneous. Next let $\widetilde{p},\widetilde{q} \in \widetilde{M}$ be such that $\sigma(\widetilde{p}) = p$, $\sigma(\widetilde{q}) = q$, where σ denotes the projection of $\widetilde{M}$ on M. There exists an isometry $\widetilde{\varphi}$ of $\widetilde{M}$ such that $\widetilde{\varphi}(\widetilde{p}) = \widetilde{q}$ and this isometry induces a local isometry φ of M such that $\varphi(p) = q$. (See [56] for more details about the covering spaces of a homogeneous manifold.)

DEFINITION 2.2. Let T be a homogeneous structure on (M,g) and T' a homogeneous structure on (M',g'). Then T and T' are said to be *isomorphic* if there exists an isometry $\varphi : (M,g) \to (M',g')$ such that

$$\varphi_*(T_X Y) = T'_{\varphi_* X}\varphi_* Y , \qquad X,Y \in \mathfrak{X}(M). \tag{2.1}$$

Note that if $\tilde{\nabla} = \nabla - T$ and $\tilde{\nabla}' = \nabla' - T'$, an isomorphism between T and T' is just an isometry of (M,g) on (M',g') which is also an *affine transformation* with respect to the connections $\tilde{\nabla}$ and $\tilde{\nabla}'$, i.e.

$$\varphi_*(\tilde{\nabla}_X Y) = \tilde{\nabla}'_{\varphi_* X}\varphi_* Y \,, \qquad X,Y \in \mathfrak{X}(M).$$

Now let T' on (M',g') be a homogeneous structure isomorphic to the homogeneous structure on (M,g) and let $p' = \varphi(p) \in M'$ where $p \in M$. Further let $\mathfrak{g} = \mathfrak{m} \oplus \mathfrak{k}$, $\mathfrak{g}' = \mathfrak{m}' \oplus \mathfrak{k}'$ be the corresponding Lie algebras constructed following the method of chapter 1, section C. Note that $\mathfrak{m} = T_pM$, $\mathfrak{m}' = T_{p'}M'$ have an inner product induced by the metric on M and M'. Then we have

THEOREM 2.3. *There exists a Lie algebra isomorphism* $\psi : \mathfrak{g} \to \mathfrak{g}'$ *such that* $\psi(\mathfrak{k}) = \mathfrak{k}'$ *and* $\psi(\mathfrak{m}) = \mathfrak{m}'$. *Further the restriction of* ψ *to* $\mathfrak{m}$ *is an isometry.*

Proof. We put

$$\psi(x) = (\varphi_*)_p(x) \,, \qquad x \in \mathfrak{m} \,, \tag{2.3}$$

$$\psi(A) = (\varphi_*)_p \circ A \circ (\varphi_*)_p^{-1} \,, \quad A \in \mathfrak{k} \,, \tag{2.4}$$

where φ denotes the isometry of (M,g) on (M',g'). Since φ is an isometry we have

$$\psi \circ (R_p)_{xy} \circ \psi^{-1} = (R'_{p'})_{\psi x \psi y} \tag{2.5}$$

where $p' = \varphi(p)$. But φ is also an isomorphism for the homogeneous structures T and T'. Hence

$$\psi \circ (T_p)_x \circ \psi^{-1} = (T'_{p'})_{\psi(x)} \,. \tag{2.6}$$

Now using (1.79) for $\mathfrak{g}$ and $\mathfrak{g}'$ we obtain easily that $\psi : \mathfrak{g} \to \mathfrak{g}'$, defined by (2.3) and (2.4), is a Lie algebra isomorphism. Further it is clear that $\psi(\mathfrak{k}) = \mathfrak{k}'$, $\psi(\mathfrak{m}) = \mathfrak{m}'$ and that $\psi_{|\mathfrak{m}} : \mathfrak{m} \to \mathfrak{m}'$ is an isometry.

In order to have a characterization of isomorphic homogeneous structures we prove the converse theorem.

<u>THEOREM 2.4.</u> *Let* (M,g) *and* (M',g') *be connected, complete and simply connected manifolds with homogeneous structure* T *on* M *and* T' *on* M'. *Further suppose that there exists a Lie algebra isomorphism* $\psi : \mathfrak{g} \to \mathfrak{g}'$ *such that* $\psi(\mathfrak{m}) = \mathfrak{m}'$, $\psi(\mathfrak{k}) = \mathfrak{k}'$ *and* $\psi_{|\mathfrak{m}} : \mathfrak{m} \to \mathfrak{m}'$ *is an isometry. Then* T *and* T' *are isomorphic homogeneous structures.*

<u>Proof.</u> Let $x,y \in \mathfrak{m} = T_pM$. Then we have

$$\psi([x,y]) = \psi(T_xy - T_yx) + \psi(\tilde{R}_{xy}).$$

Further, since ψ is a Lie algebra isomorphism, we have :

$$\psi([x,y]) = [\psi(x),\psi(y)]$$

$$= T'_{\psi(x)}\psi(y) - T'_{\psi(y)}\psi(x) + \tilde{R}'_{\psi(x)\psi(y)}.$$

Since $\psi(\mathfrak{k}) = \mathfrak{k}'$ and $\psi(\mathfrak{m}) = \mathfrak{m}'$ we obtain

$$\psi(T_xy - T_yx) = T'_{\psi(x)}\psi(y) - T'_{\psi(y)}\psi(x) , \tag{2.7}$$

$$\psi(\tilde{R}_{xy}) = \tilde{R}'_{\psi(x)\psi(y)}. \tag{2.8}$$

Hence $\psi_{|\mathfrak{m}}$ is an isometry which preserves the curvature. Further (2.7) shows that $\psi_{|\mathfrak{m}}$ also preserves the torsion $\tilde{S}$. So, using a standard argument, we may conclude that there exists a transformation $\varphi : M \to M'$ which is an affine transformation with respect to the connections $\tilde{\nabla}$, $\tilde{\nabla}'$, such that $\varphi(p) = p'$ and $(\varphi_*)_p = \psi$. (See [26, vol. I, p. 265]. Here we note that $\tilde{\nabla}$ and $\tilde{\nabla}'$ are complete since these connections are metric. (Proposition 1.5)) Hence φ is an isometry because it is an affine transformation with respect to $\tilde{\nabla}$ and $\tilde{\nabla}'$ which is an isometry at p. (See Proposition 1.3 and the note after it.)

<u>REMARKS.</u>

A. With the hypothesis of Theorem 2.4 we may conclude that (M,g) and (M',g') are homogeneous. The groups G and G' acting on M and M' as isometry groups and constructed from T and T' are locally isomorphic and isomorphic if they are simply connected. The same is true for the two isotropy groups.

B. Let T be a homogeneous structure on (M,g) and φ an isometry of (M,g). Then

$$T'_X Y = \varphi_* T_{\varphi_*^{-1} X} \varphi_*^{-1} Y \quad , \quad X,Y \in \mathfrak{X}(M) \; ,$$

is also a homogeneous structure on (M,g) since it is also a solution of the system (AS). In general T' is different from T and hence T' gives another solution of that system (see chapter 4 for the Poincaré half-plane).

There are also examples where T' = T for all isometries of (M,g). See for example the case of the Heisenberg group in chapter 7.

This proves that in general the system (AS) does not admit a *unique* solution. But the existence of isomorphic homogeneous structures does not explain completely the existence of several solutions of the system (AS). Indeed, we may have the following two situations :

1. There exist two homogeneous structures T_1 and T_2 which are not isomorphic but which give rise to the same Lie algebra $\mathfrak{g}$ with different decompositions : $\mathfrak{g} = \mathfrak{m}_1 \oplus \mathfrak{k}_1 = \mathfrak{m}_2 \oplus \mathfrak{k}_2$. This means that there does not exist a Lie algebra isomorphism of $\mathfrak{g}$ onto itself such that $\psi(\mathfrak{m}_1) = \mathfrak{m}_2$, $\psi(\mathfrak{k}_1) = \mathfrak{k}_2$ and $\psi_{|\mathfrak{m}_1} : \mathfrak{m}_1 \to \mathfrak{m}_2$ is an isometry. The Heisenberg group provides an example for this situation (see chapter 7).

2. There exist two homogeneous structures T_1 and T_2 with non-isomorphic Lie algebras $\mathfrak{g}_1$, $\mathfrak{g}_2$. S^6 is an example for this situation. (See chapters 7 and 8 for other examples.) This means that we have different representations of the homogeneous space as a quotient space G/K. In the case of the sphere S^6 we have for example $S^6 = SO(7)/SO(6)$ and $S^6 = G_2/SU(3)$.

3. THE EIGHT CLASSES OF HOMOGENEOUS STRUCTURES

Let p be a point of M and $V = T_pM$. V is a Euclidean vector space over $\mathbb{R}$ with inner product $<,>$ induced from the metric g on M. Instead of considering the tensors T of type (1,2) we prefer to work here with those of type (0,3) given by the isomorphism

$$T_{xyz} = < T_x y, z > , \qquad x,y,z \in V. \tag{3.1}$$

Next we consider the vector space $\mathcal{T}(V)$, subspace of $\overset{3}{\otimes} V^*$, determined by all the (0,3)-tensors having the same symmetries as a homogeneous structure, i.e.

$$\mathcal{T}(V) = \{T \in \overset{3}{\otimes} V^* | T_{xyz} = - T_{xzy} , \quad x,y,z \in V\} , \tag{3.2}$$

where V^* denotes the dual vector space of V. $\mathcal{T}(V)$ is a Euclidean vector space with inner product defined by

$$< T,T' > = \sum_{i,j,k} T_{e_i e_j e_k} T'_{e_i e_j e_k} , \tag{3.3}$$

where $(e_1,\dots,e_n)$ is an arbitrary orthonormal basis of V.

Further there is a natural action of the orthogonal group O(V) on $\mathcal{T}(V)$ determined by

$$(aT)_{xyz} = T_{a^{-1}x\, a^{-1}y\, a^{-1}z} , \tag{3.4}$$

for $x,y,z \in V$ and $a \in O(V)$.

Next we determine the decomposition of $\mathcal{T}(V)$ into irreducible invariant components under this action of the orthogonal group. Let

$$c_{12}(T)(z) = \sum_i T_{e_i e_i z} , \qquad z \in V , \tag{3.5}$$

for an arbitrary orthonormal basis (e_i) of V and put

$$\mathcal{T}_1(V) = \{T \in \mathcal{T}(V) \mid T_{xyz} = \langle x,y \rangle \varphi(z) - \langle x,z \rangle \varphi(y) , \ \varphi \in V^*\}, \tag{3.6}$$

$$\mathcal{T}_2(V) = \{T \in \mathcal{T}(V) \mid \underset{x,y,z}{\mathfrak{S}} T_{xyz} = 0 , \ c_{12}(T) = 0\} , \tag{3.7}$$

$$\mathcal{T}_3(V) = \{T \in \mathcal{T}(V) \mid T_{xyz} + T_{yxz} = 0\} \tag{3.8}$$

where $x,y,z \in V$. Here $\underset{x,y,z}{\mathfrak{S}}$ denotes the cyclic sum with respect to x,y and z.

We have

THEOREM 3.1. *Let* $\dim V \geq 3$. *Then* $\mathcal{T}(V)$ *is the orthogonal direct sum of the subspaces* $\mathcal{T}_i(V)$, $i = 1,2,3$. *Moreover, these spaces are invariant and irreducible under the action of* $O(V)$.

Let $\dim V = 2$. *Then we have* $\mathcal{T}(V) = \mathcal{T}_1(V)$ *where* $\mathcal{T}(V)$ *is irreducible.*

Proof. The theorem follows from the results of [53] or the theorem may be proved using the following remarks. First, it is easily seen by direct calculation that the spaces $\mathcal{T}_i(V)$ are invariant and orthogonal. Next we note that the space of *quadratic invariants* of T has dimension 3 for $\dim V > 2$ and is generated by

$$\begin{cases} \|T\|^2 = \sum_{i,j,k} T^2_{e_i e_j e_k} , \\ \langle T,\hat{T} \rangle = \sum_{i,j,k} T_{e_i e_j e_k} T_{e_j e_i e_k} , \\ \|c_{12}(T)\|^2 = \sum_{i,j,k} T_{e_i e_i e_k} T_{e_j e_j e_k} . \end{cases} \tag{3.9}$$

This implies the irreducibility of $\mathcal{T}_i(V)$ in that case (see also [6],[15], [49]).

Finally it is easy to see that

$$\dim \mathcal{T}(V) = \frac{n^2(n-1)}{2} \tag{3.10}$$

and

$$\begin{cases} \dim \mathcal{T}_1(V) = n\ , \\ \dim \mathcal{T}_2(V) = \dfrac{n(n-2)(n+2)}{3}\ , \\ \dim \mathcal{T}_3(V) = \binom{n}{3}. \end{cases} \tag{3.11}$$

Hence, for n = 2, we have $\mathcal{T}(V) = \mathcal{T}_1(V)$ and $\mathcal{T}(V)$ is isomorphic to V^* which is irreducible.

Note that

$$\mathcal{T}_2 \oplus \mathcal{T}_3 = \ker c_{12}.$$

Further $\mathcal{T}_1(V)$ is always isomorphic to V^* and $\mathcal{T}_3(V)$ is isomorphic to $\Lambda^3 V^*$, the space of alternating 3-forms on V.

The following theorem is easy to prove.

THEOREM 3.2. *Let* $T \in \mathcal{T}(V)$ *and denote by* $p_i(T) = T_i$, $i = 1,2,3$, *the projections of* T *on* $\mathcal{T}_i(V)$. *Then we have*

$$p_1(T)_{xyz} = \langle x,y \rangle \varphi(z) - \langle x,z \rangle \varphi(y) \tag{3.12}$$

where

$$\varphi(z) = \frac{1}{n-1} c_{12}(T)(z)\ , \tag{3.13}$$

$$p_3(T)_{xyz} = \frac{1}{3} \underset{xyz}{\mathfrak{S}}\, T_{xyz}\ , \tag{3.14}$$

and

$$p_2(T) = T - p_1(T) - p_3(T)\ , \tag{3.15}$$

with $x,y,z \in V$.

Finally we give expressions for the squares of the lengths of the projections as functions of the quadratic invariants. We have

THEOREM 3.3. *Let* $p_i(T)$ *denote the projections of* T. *Then*

$$\|p_1(T)\|^2 = \frac{2}{n-1} \|c_{12}(T)\|^2 , \tag{3.16}$$

$$\|p_2(T)\|^2 = \frac{2}{3} \{\|T\|^2 + < T,\hat{T} >\} - \frac{2}{n-1} \|c_{12}(T)\|^2 , \tag{3.17}$$

$$\|p_3(T)\|^2 = \frac{1}{3} \{\|T\|^2 - 2 < T,\hat{T} >\} . \tag{3.18}$$

Proof. By direct calculation using Theorem 3.2.

It follows from Theorem 3.1 that there are, in general, eight invariant subspaces of $\mathcal{T}(V)$, including the trivial spaces. This suggests the following.

DEFINITION 3.4. Let $\mathcal{J}(V)$ be an invariant subspace of $\mathcal{T}(V)$. We say that a homogeneous structure T on (M,g) is of *type* $\mathcal{J}$ when $T_p \in \mathcal{J}(T_pM)$ for all $p \in M$.

Note that every class $\mathcal{J}$ of the eight classes of homogeneous structures is invariant under isomorphisms of homogeneous structures.

We have from Theorem 3.3 :

THEOREM 3.5. *Let* T *be a homogeneous structure. Then*

(i) T *is of type* $\mathcal{T}_1$ *if and only if* $\|T\|^2 = 2 < T,\hat{T} > = \frac{2}{n-1} \|c_{12}(T)\|^2$;

(ii) T *is of type* $\mathcal{T}_2$ *if and only if* $\|c_{12}(T)\|^2 = 0$ *and* $\|T\|^2 = 2 < T,\hat{T} >$;

(iii) T *is of type* $\mathcal{T}_3$ *if and only if* $\|T\|^2 + < T,\hat{T} > = 0$;

(iv) T *is of type* $\mathcal{T}_1 \oplus \mathcal{T}_2$ *if and only if* $\|T\|^2 - 2 < T,\hat{T} > = 0$;

(v) T *is of type* $\mathcal{T}_1 \oplus \mathcal{T}_3$ *if and only if* $\|T\|^2 + < T,\hat{T} > = \frac{3}{n-1} \|c_{12}(T)\|^2$;

(vi) T *is of type* $\mathcal{T}_2 \oplus \mathcal{T}_3$ *if and only if* $\|c_{12}(T)\|^2 = 0$.

Finally in table I we give a résumé of the characterizing identities for the eight classes. All these identities are trivial except perhaps that for the class $\mathcal{C}_1 \oplus \mathcal{C}_3$. To obtain this we note that if $T \in \mathcal{C}_1(V) \oplus \mathcal{C}_3(V)$, we have

$$T_{xyz} = \langle x,y \rangle \psi(z) - \langle x,z \rangle \psi(y) + (T_3)_{xyz}$$

for all $x,y,z \in V$. Hence we obtain

$$T_{xyz} + T_{yxz} = 2 \langle x,y \rangle \psi(z) - \langle x,z \rangle \psi(y) - \langle y,z \rangle \psi(x) \quad (3.19)$$

Classes	Defining conditions
Symmetric structures	$T = 0$
$\mathcal{C}_1$	$T_{XYZ} = g(X,Y)\varphi(Z) - g(X,Z)\varphi(Y) \quad , \quad \varphi \in \Lambda^1 M$
$\mathcal{C}_2$	$\underset{X,Y,Z}{\mathfrak{S}} T_{XYZ} = 0 \ , \ c_{12}(T) = 0$
$\mathcal{C}_3$	$T_{XYZ} + T_{YXZ} = 0$
$\mathcal{C}_1 \oplus \mathcal{C}_2$	$\underset{X,Y,Z}{\mathfrak{S}} T_{XYZ} = 0$
$\mathcal{C}_1 \oplus \mathcal{C}_3$	$T_{XYZ} + T_{YXZ} = 2g(X,Y)\varphi(Z) - g(X,Z)\varphi(Y) - g(Y,Z)\varphi(X) \quad , \quad \varphi \in \Lambda^1 M$
$\mathcal{C}_2 \oplus \mathcal{C}_3$	$c_{12}(T) = 0$
$\mathcal{C} = \mathcal{C}_1 \oplus \mathcal{C}_2 \oplus \mathcal{C}_3$	no conditions

TABLE I.

since $(T_3)_{xyz} + (T_3)_{yxz} = 0$ by definition. Conversely, when T satisfies (3.19) we obtain by contraction

$$c_{12}(T)(z) = (n-1)\psi(z).$$

Hence

$$(T_1)_{xyz} = \langle x,y \rangle \psi(z) - \langle x,z \rangle \psi(y)$$

and

$$(T_2)_{xyz} = T_{xyz} - (T_1)_{xyz} - (T_3)_{xyz} .$$

So

$$(T_2)_{xyz} + (T_2)_{yxz} = 0.$$

This implies $T_2 \in \mathcal{T}_2(V) \cap \mathcal{T}_3(V)$ and hence $T_2 = 0$.

We refer to chapters 4-8 for examples of Riemannian manifolds with a homogeneous structure which belongs to one of these eight classes.

4. HOMOGENEOUS STRUCTURES ON SURFACES

In this chapter we always suppose that (M,g) is a two-dimensional Riemannian manifold. We shall determine all the homogeneous structures on a homogeneous surface.

Since $\mathcal{T}(V) = \mathcal{T}_1(V)$ for dim V = 2, all homogeneous structures must belong to $\mathcal{T}_1$ and hence we can write

$$T_X Y = g(X,Y)\xi - g(\xi,Y)X \quad , \tag{4.1}$$

where $X, Y, \xi \in \mathfrak{X}(M)$.

First we prove

<u>THEOREM 4.1.</u> *Let* T *be a nonzero homogeneous structure on the connected surface* (M,g). *Then* (M,g) *has constant negative curvature.*

<u>Proof</u>. We put $\tilde{\nabla} = \nabla - T$ where T is given by (4.1). Then $\tilde{\nabla}T = 0$ if and only if $\tilde{\nabla}\xi = 0$ since $\tilde{\nabla}$ is a metric connection. Hence $g(\xi,\xi)$ is a non-vanishing constant if $T \neq 0$. Further (4.1) and $\tilde{\nabla}\xi = 0$ imply

$$\nabla_X \xi = g(X,\xi)\xi - g(\xi,\xi)X \tag{4.2}$$

and hence

$$\begin{aligned} R_{XY}\xi &= \nabla_{[X,Y]}\xi - [\nabla_X, \nabla_Y]\xi \\ &= -c^2\{g(X,\xi)Y - g(Y,\xi)X\} \quad , \end{aligned} \tag{4.3}$$

where $g(\xi,\xi) = c^2$. This proves the theorem.

COROLLARY 4.2. *The only homogeneous structure on* $\mathbb{R}^2$ *and* S^2 *is given by* $T = 0$.

Note that the condition given in Theorem 4.1 is only a necessary condition which is not sufficient. Indeed, we proved that if there exists a homogeneous structure $T \neq 0$, then there must exist a non-vanishing vector field of constant length on M. When M is compact the Euler-Poincaré characteristic gives an obstruction to the existence of such a structure. This implies that a compact surface of negative curvature does not admit a homogeneous structure $T \neq 0$. Of course we always have the structure $T = 0$. These compact spaces are locally homogeneous but not homogeneous since their full isometry group is finite (there are no nonvanishing Killing vector fields) [25].

If M is simply connected we have

THEOREM 4.3. *Let* (M,g) *be a connected, complete and simply connected surface. Then* (M,g) *admits a homogeneous structure* $T \neq 0$ *if and only if* (M,g) *is isometric to the hyperbolic plane.*

Proof. Necessity follows at once from Theorem 4.1.

To prove sufficiency we consider the Poincaré half-plane $\mathbb{H}^2 = \{(y^1,y^2), y^1 > 0\}$ with the metric

$$g = r^2(y^1)^{-2}\{(dy^1)^2 + (dy^2)^2\} , \qquad (4.4)$$

where r is a constant. The Gauss curvature K of $\mathbb{H}^2$ is given by $K = -r^{-2}$.

Consider the structure T given by (4.1) where

$$\xi = \frac{1}{r^2}\, y^1 \frac{\partial}{\partial y^1} . \qquad (4.5)$$

It is easy to see that ξ satisfies (4.2) and hence $\tilde{\nabla}T = 0$. So this $T \neq 0$ determines a homogeneous structure since $\tilde{\nabla}R = 0$ is automatically satisfied for all Riemannian manifolds of constant curvature.

For the hyperbolic plane it is possible to determine *all* homogeneous structures T. To do this we have to find all the solutions of (4.2). In the case of the hyperbolic plane (4.2) is equivalent to

$$\begin{cases} \frac{1}{r} y^1 \frac{\partial \xi^1}{\partial y^1} = -(\xi^2)^2 , \\ \frac{1}{r} y^1 \frac{\partial \xi^2}{\partial y^1} = \xi^1 \xi^2 , \\ \frac{1}{r} y^1 \frac{\partial \xi^1}{\partial y^2} = -\frac{1}{r} \xi^2 + \xi^1 \xi^2 , \\ \frac{1}{r} y^1 \frac{\partial \xi^2}{\partial y^2} = \frac{1}{r} \xi^1 - (\xi^1)^2 , \end{cases} \tag{4.6}$$

where

$$\xi = \xi^1 E_1 + \xi^2 E_2$$

and

$$E_1 = \frac{1}{r} y^1 \frac{\partial}{\partial y^1} , \quad E_2 = \frac{1}{r} y^1 \frac{\partial}{\partial y^2} . \tag{4.7}$$

(E_1, E_2) is an orthonormal basis for the hyperbolic plane.

The integrability conditions of the system (4.6) can be reduced to

$$(\xi^1)^2 + (\xi^2)^2 = \frac{1}{r^2} .$$

Further an easy calculation shows that the solutions of (4.6) are

$$\xi^1 = \frac{1}{r} , \quad \xi^2 = 0 \tag{4.8}$$

and

$$\begin{cases} \xi^1 = \frac{1}{r} \frac{(\lambda - y^2)^2 - (y^1)^2}{(\lambda - y^2)^2 + (y^1)^2} , \\ \xi^2 = \frac{2}{r} \frac{(\lambda - y^2) y^1}{(\lambda - y^2)^2 + (y^1)^2} , \end{cases} \tag{4.9}$$

where λ is an arbitrary constant.

We may conclude that there are an *infinite* number of homogeneous structures on the hyperbolic plane but we have the following

THEOREM 4.4. *Up to isomorphism* $\mathbb{H}^2$ *has only two homogeneous structures, namely* $T = 0$ *and* $T_X Y = g(X,Y)\xi - g(\xi,Y)X$ *where* ξ *is given by* (4.5) *and* $X,Y \in \mathfrak{X}(\mathbb{H}^2)$.

Proof. The full isometry group of $\mathbb{H}^2$ is the group

$$SL(2,\mathbb{R})/\pm I \cup \alpha(SL(2,\mathbb{R})/\pm I)$$

where α is the isometry of $\mathbb{H}^2$ determined by

$$\alpha(y^1,y^2) = (y^1,-y^2). \tag{4.10}$$

(See [56].) We recall that the action of $SL(2,\mathbb{R})/\pm I$, the connected component of the identity, on $\mathbb{H}^2$ is given by

$$z = y^2 + iy^1 \mapsto \frac{az+b}{cz+d} \tag{4.11}$$

where

$$\begin{pmatrix} a & b \\ c & d \end{pmatrix} \in SL(2,\mathbb{R}) .$$

The dual frame (θ^1,θ^2) of the frame (E_1,E_2), given by (4.7), is determined by

$$\theta^1 = r\,\frac{dy^1}{y^1} \;, \quad \theta^2 = r\,\frac{dy^2}{y^1} . \tag{4.12}$$

First we have

$$\alpha^*\theta^1 = \theta^1 \;, \; \alpha^*\theta^2 = -\theta^2 . \tag{4.13}$$

Next let L_A denote the left translation of $\mathbb{H}^2$ induced by

$$A = \begin{pmatrix} a & b \\ c & d \end{pmatrix} \in SL(2,\mathbb{R}) .$$

Then we have

$$L_A^*\theta^1 = \frac{|cz + d|^2}{2i}\left\{\frac{\theta^2 + i\theta^1}{(cz+d)^2} - \frac{\theta^2 - i\theta^1}{(c\bar{z}+d)^2}\right\}, \tag{4.14}$$

$$L_A^*\theta^2 = \frac{|cz + d|^2}{2}\left\{\frac{\theta^2 + i\theta^1}{(cz+d)^2} + \frac{\theta^2 - i\theta^1}{(c\bar{z}+d)^2}\right\}. \tag{4.15}$$

Hence, if ξ is the vector field given by (4.5), we have

$$\alpha_*\xi = \xi , \tag{4.16}$$

and

$$(L_A)_*\xi = \frac{(cy^2 + d)^2 - c^2(y^1)^2}{r\{(cy^2+d)^2 + c^2(y^1)^2\}} E_1 + 2\frac{c(d - cy^2)y^1}{r\{(cy^2+d)^2 + c^2(y^1)^2\}} E_2. \tag{4.17}$$

For $c = 0$ we have $(L_A)_*\xi = \xi$ and for $c \neq 0$ we obtain (4.9) by putting $\lambda = -\frac{d}{c}$. This implies the required result.

Note that the first homogeneous structure, $T = 0$, corresponds to the symmetric case, i.e. to the representation of $\mathbb{H}^2$ as

$$\mathbb{H}^2 = SO_o(1,2)/SO(2). \tag{4.18}$$

where $SO_o(1,2) = SL(2,R)/\pm I$ is the connected component of the identity of the *Lorentz group*.

The second homogeneous structure corresponds to the representation of $\mathbb{H}^2$ as a Lie group with a product defined by

$$(y_1,y_2)(y_1',y_2') = (y_1y_1',y_1y_2' + y_2). \tag{4.19}$$

This is a semi-direct product of the multiplicative group $\mathbb{R}_0^+$ and the additive group $\mathbb{R}$. To verify this statement it is sufficient to compute the Lie algebra starting from T and using (1.79). Putting $\tilde{\nabla} = \nabla - T$ one finds at once that $\tilde{\nabla}$ is flat and hence the holonomy algebra $\mathfrak{k}$ of $\tilde{\nabla}$ is trivial. So we have $\mathfrak{g} = \mathfrak{m} = T_pM$ with the following Lie bracket :

$$[\xi_1,\xi_2] = T_{\xi_1}\xi_2 - T_{\xi_2}\xi_1.$$

Now let (e_1,e_2) be an orthonormal basis of m such that $e_1 = r\xi$. Then we obtain

$$[e_1,e_2] = \frac{1}{r}\, e_1 \,. \tag{4.20}$$

Hence g is the semi-direct sum of two one-dimensional Lie algebras. It is not difficult to check that g is the Lie algebra of $\mathbb{H}^2$ with the product (4.19).

It is possible to obtain a confirmation and in some sense an explanation of this result by considering the *Iwasawa decomposition* of $SL(2,\mathbb{R})$ (see [18]). Indeed we have

$$SL(2,\mathbb{R}) = KAN \tag{4.21}$$

where K,A,N are the subgroups of $SL(2,\mathbb{R})$ defined as follows :

$$K = SO(2)\ ,\ A = \left\{\begin{pmatrix} u & 0 \\ 0 & \frac{1}{u}\end{pmatrix},\ u > 0\right\}\ ,\ N = \left\{\begin{pmatrix} 1 & v \\ 0 & 1\end{pmatrix},\ v \in \mathbb{R}\right\}. \tag{4.22}$$

Then

$$\varphi : AN \to \mathbb{H}^2 : \begin{pmatrix} u & v \\ 0 & \frac{1}{u}\end{pmatrix} \mapsto (u^2, uv)$$

defines an isomorphism of AN onto the Lie group $\mathbb{H}^2$ with the product (4.19). AN acts effectively, transitively and simply on $\mathbb{H}^2$ by (4.11). Finally we note that the elements of $SL(2,\mathbb{R})$ for which the nonzero homogeneous structure is invariant are those of type

$$\begin{pmatrix} a & b \\ 0 & d\end{pmatrix}$$

(see (4.17)) and these are exactly the elements of AN. So the set of ∞^1 homogeneous structures $T \neq 0$ on $\mathbb{H}^2$ is parametrized by $SO(2)$.

5. HOMOGENEOUS STRUCTURES OF TYPE $\mathcal{T}_1$

Let (M,g) be a connected Riemannian manifold of dimension n and suppose M admits a nontrivial homogeneous structure T of type $\mathcal{T}_1$, i.e. there exists a tensor field T on M given by

$$T_{XYZ} = g(X,Y)\varphi(Z) - g(X,Z)\varphi(Y) ,$$

where $X,Y,Z \in \mathfrak{X}(M)$ and φ is a nonzero 1-form on M. Equivalently, if

$$g(\xi,X) = \varphi(X) , \qquad X \in \mathfrak{X}(M) ,$$

T is given by

$$T_X Y = g(X,Y)\xi - g(\xi,Y)X , \qquad X,Y \in \mathfrak{X}(M).$$

In chapter 4 we have seen that all the nontrivial homogeneous structures on a surface are of this type. The main purpose of this chapter is to consider the case $n > 2$ and to show that Theorem 4.1 and Theorem 4.2 are still valid.

<u>THEOREM 5.1.</u> *Let* (M,g) *be a connected Riemannian manifold which admits a homogeneous structure* $T \neq 0$ *of type* $\mathcal{T}_1$. *Then* M *has constant negative curvature.*

<u>THEOREM 5.2.</u> *Let* (M,g) *be a connected, complete and simply connected Riemannian manifold. Then* (M,g) *admits a nonvanishing homogeneous structure* $T \in \mathcal{T}_1$ *if and only if* (M,g) *is isometric to the hyperbolic space.*

To prove these theorems we establish several lemmas.

LEMMA 5.3. *Let $\mathfrak{D}_1$ be the one-dimensional distribution of* M *generated by* ξ *and let $\mathfrak{D}_2$ be the* (n-1)*-dimensional distribution orthogonal to $\mathfrak{D}_1$. Then $\mathfrak{D}_2$ is involutive.*

Proof. Put $\tilde{\nabla} = \nabla - T$. Then $\tilde{\nabla}T = 0$ if and only if

$$\tilde{\nabla}\xi = 0 \tag{5.1}$$

since $\tilde{\nabla}$ is a metric connection. Hence $g(\xi,\xi) = c^2$ is constant. Further $c \neq 0$ since $T \neq 0$.

Next, (5.1) is equivalent to

$$\nabla_X\xi = g(X,\xi)\xi - g(\xi,\xi)X , \qquad X \in \mathfrak{X}(M). \tag{5.2}$$

Hence

$$\nabla_X\xi = - c^2 X \tag{5.3}$$

if $X \in \mathfrak{D}_2$.

Finally, let $X,Y \in \mathfrak{D}_2$. Then

$$g([X,Y],\xi) = g(\nabla_X Y,\xi) - g(\nabla_Y X,\xi)$$

$$= - g(Y,\nabla_X\xi) + g(X,\nabla_Y\xi) = 0.$$

This means that $\mathfrak{D}_2$ is involutive.

LEMMA 5.4. *Let M_2 be a maximal integral manifold of $\mathfrak{D}_2$. Then the induced metric on M_2 is locally symmetric.*

Proof. The Riemannian connection ∇' on M_2 is defined by

$$\nabla'_X Y = \nabla_X Y - g(\nabla_X Y,\zeta)\zeta , \tag{5.4}$$

where $X,Y \in \mathfrak{X}(M_2)$ and ζ is a unit normal vector field. Hence $\xi = c\zeta$.

Using (5.3), (5.4) becomes

$$\nabla'_X Y = \nabla_X Y - g(X,Y)\xi . \tag{5.5}$$

Further, since $T_X Y = g(X,Y)\xi$ for $X,Y \in \mathfrak{X}(M_2)$, we get

$$\nabla'_X Y = \tilde{\nabla}_X Y , \quad X,Y \in \mathfrak{X}(M_2). \tag{5.6}$$

Hence M_2 is autoparallel with respect to $\tilde{\nabla}$. Also it follows from (5.5) that the second fundamental form α of M_2 is given by

$$\alpha(X,Y) = -g(X,Y)\xi . \tag{5.7}$$

This means that M_2 is a totally umbilical submanifold with mean curvature vector $H = -\xi$.

Now we use the Gauss equation to compute the curvature tensor R' of ∇' on M_2. With (5.7) we obtain

$$R'_{XYZW} = R_{XYZW} + c^2\{g(X,Z)g(Y,W) - g(X,W)g(Y,Z)\} , \tag{5.8}$$

$X,Y,Z,W \in \mathfrak{X}(M_2)$. From this we obtain

$$\nabla' R' = \nabla'(R_{|M_2}). \tag{5.9}$$

But since M_2 is autoparallel with respect to $\tilde{\nabla}$, we also have

$$\nabla'(R_{|M_2}) = (\tilde{\nabla}R)_{|M_2} \tag{5.10}$$

and hence, since $\tilde{\nabla}R = 0$, we obtain the required result $\nabla' R' = 0$.

<u>LEMMA 5.5.</u> (M,g) *is locally isometric to the product* $\mathbb{R} \times M_2$ *with the metric*

$$ds^2 = c^2 dt^2 + e^{-2c^2 t} g_2 ,$$

where g_2 *denotes the induced locally symmetric metric on* M_2.

<u>Proof</u>. Let $p \in M$ and denote by M_1 the integral manifold of $\mathcal{D}_1$ through p and by M_2 the maximal integral manifold of $\mathcal{D}_2$ through p. Further let $(t,x^1,\ldots,x^{n-1})$ be a chart with domain $\mathcal{U} \subset M$ centered at p (i.e. $(t,x^1,\ldots,x^{n-1})(p) = (0,\ldots,0)$) such that

$$\xi_{|\mathcal{U}} = \frac{\partial}{\partial t} \tag{5.11}$$

and $(\frac{\partial}{\partial x^i}, i = 1,\ldots,n-1)$ is a local basis for $\mathcal{D}_2$. With respect to this chart the connected component $\mathcal{C}_1$ of $M_1 \cap \mathcal{U}$ which contains p is given by

$$\mathcal{C}_1 = \{(t,0,\ldots,0)\}$$

and the connected component $\mathcal{C}_2$ of $M_2 \cap \mathcal{U}$ containing p by

$$\mathcal{C}_2 = \{(0,x^1,\ldots,x^{n-1})\}.$$

Locally, the metric g of M can be written in the following way :

$$g = c^2 dt^2 + \sum_{i,j=1}^{n-1} g_{ij}(t,x)\ dx^i \otimes dx^j$$

since

$$g(\frac{\partial}{\partial x^i}, \frac{\partial}{\partial t}) = 0 \ , \ g(\frac{\partial}{\partial t}, \frac{\partial}{\partial t}) = c^2.$$

On the other hand we have

$$\frac{\partial}{\partial t} g_{ij}(t,x) = \frac{\partial}{\partial t} g(\frac{\partial}{\partial x^i}, \frac{\partial}{\partial x^j})(t,x)$$

$$= g(\nabla_{\frac{\partial}{\partial t}} \frac{\partial}{\partial x^i}, \frac{\partial}{\partial x^j}) + g(\frac{\partial}{\partial x^i}, \nabla_{\frac{\partial}{\partial t}} \frac{\partial}{\partial x^j})$$

$$= g(\nabla_{\frac{\partial}{\partial x^i}} \frac{\partial}{\partial t}, \frac{\partial}{\partial x^j}) + g(\frac{\partial}{\partial x^i}, \nabla_{\frac{\partial}{\partial x^j}} \frac{\partial}{\partial t}).$$

Using (5.11) and (5.3) we obtain

$$\frac{\partial}{\partial t} g_{ij}(t,x) = - 2c^2 g_{ij}(t,x). \tag{5.12}$$

This implies

$$g_{ij}(t,x) = e^{-2c^2 t} \tilde{g}_{ij}(x). \tag{5.13}$$

Note that

$$g_2\left(\frac{\partial}{\partial x^i}, \frac{\partial}{\partial x^j}\right)(x) = g\left(\frac{\partial}{\partial x^i}, \frac{\partial}{\partial x^j}\right)(0,x) = g_{ij}(0,x) = \tilde{g}_{ij}(x).$$

This proves the required result.

LEMMA 5.6. *The metric* g_2 *on* $\mathfrak{D}_2$ *is flat.*

Proof. Using Lemma 5.5 we can write R and ∇R as functions of the curvature tensor R' of M_2.

First we note that

$$\nabla_\xi \xi = 0 \ , \tag{5.14}$$

$$\nabla_X \xi = \nabla_\xi X = - c^2 X \ , \tag{5.15}$$

$$\nabla_X Y = g(X,Y)\xi + \tilde{\nabla}_X Y, \tag{5.16}$$

where $X,Y \in \mathfrak{X}(M_2)$. Hence we obtain

$$R_{\xi Y}\xi = - c^4 Y \ , \tag{5.17}$$

$$R_{\xi Y}Z = c^2 g(Y,Z)\xi \ , \tag{5.18}$$

$$R_{XY}\xi = 0 \ , \tag{5.19}$$

$$R_{XY}Z = R'_{XY}Z - c^2\{g(X,Z)Y - g(Y,Z)X\} \ , \tag{5.20}$$

where $X, Y, Z \in \mathfrak{X}(M_2)$. This implies

$$(\nabla_\xi R)_{XY}Z = 2c^2 R'_{XY}Z . \tag{5.21}$$

Next $\tilde{\nabla}R = 0$ implies

$$(\nabla_\xi R)_{XY}Z = [T_\xi, R_{XY}Z] - R_{T_\xi XY}Z - R_{XT_\xi Y}Z. \tag{5.22}$$

Since $X, Y, R_{XY}Z \in \mathfrak{X}(M_2)$ we have

$$T_\xi X = T_\xi Y = [T_\xi, R_{XY}Z] = 0 .$$

So (5.22) is satisfied if and only if $R' = 0$ since $c \neq 0$ in (5.21).

Now we are in a position to prove the theorems.

Proof of Theorem 5.1. It follows from Lemmas 5.5 and 5.6 that locally we can write

$$ds^2 = c^2 dt^2 + e^{-2c^2 t} \sum_{i=1}^{n-1} (dx^i)^2. \tag{5.23}$$

Using the transformation

$$\begin{cases} y^1 = \dfrac{1}{c^2} e^{c^2 t} , \\ y^i = cx^{i-1} , \quad i = 2, \ldots, n , \end{cases} \tag{5.24}$$

(5.23) becomes

$$ds^2 = (cy^1)^{-2} \sum_{j=1}^{n} (dy^j)^2 , \quad y^1 > 0. \tag{5.25}$$

This is the metric of the Poincaré half-space $\mathbb{H}^n$.

Proof of Theorem 5.2. The first part follows from Theorem 5.1.

Next let $M = \mathbb{H}^n$. Then it is easy to check that the vector field

$$\xi = c^2 y^1 \frac{\partial}{\partial y^1} \tag{5.26}$$

is a solution of (5.2) and hence it determines a homogeneous structure $T \neq 0$ on $\mathbb{H}^n$ since the condition $\tilde{\nabla}R = 0$ is always fulfilled for a space of constant curvature.

Using the same method as in chapter 4 it is not difficult to check that this homogeneous structure $T \in \mathcal{T}_1$ corresponds to the representation of $\mathbb{H}^n$ as the Lie group with the following product

$$(x^1,\ldots,x^n)(y^1,\ldots,y^n) = (x^1y^1, x^1y^2 + x^2,\ldots,x^1y^n + x^n). \tag{5.27}$$

So we have a solvable Lie group which is a semi-direct product of the multiplicative group $\mathbb{R}_0^+$ and the additive group $\mathbb{R}^n$.

This result is a special case of a much more general theorem. Indeed, it is known [3],[4],[17],[55] that when (M,g) is a connected, simply connected, homogeneous Riemannian manifold with nonpositive sectional curvature, then there exists a solvable Lie group G which acts transitively and *simply* on M as a group of isometries of M. Hence M can be identified with G.

Finally we note that the system (AS) for $\mathbb{H}^n$ is much more difficult to solve for $n \geqslant 3$. We do not have information about all the possible solutions. Here we only have shown that for $\mathbb{H}^n$ there are at least two nonisomorphic homogeneous structures of type $\mathcal{T}_1$ since $T = 0$ is a trivial solution. We refer to chapter 8 for the special case of $\mathbb{H}^4$.

6. NATURALLY REDUCTIVE HOMOGENEOUS SPACES AND HOMOGENEOUS STRUCTURES OF TYPE $\mathcal{T}_3$

In this chapter we give a characterization of the manifolds which admit a homogeneous structure of type $\mathcal{T}_3$. Moreover we determine all the three-dimensional manifolds with such a structure. To do this we first need to consider naturally reductive spaces.

As before let (M,g) be a connected n-dimensional Riemannian manifold. Further let M = G/K, where G is a group of isometries of M acting transitively and effectively on the manifold. We denote by K the isotropy subgroup at a point p of M. Next let $\mathfrak{g}$ denote the Lie algebra of G and $\mathfrak{k}$ the Lie algebra of K. Then M = G/K is said to be *naturally reductive* if there exists a vector subspace $\mathfrak{m}$ of $\mathfrak{g}$ such that

$$\mathfrak{g} = \mathfrak{m} \oplus \mathfrak{k} \,, \tag{6.1}$$

$$\mathrm{Ad}(K)\mathfrak{m} \subseteq \mathfrak{m}, \tag{6.2}$$

$$< [X,Y]_{\mathfrak{m}}, Z > + < [X,Z]_{\mathfrak{m}}, Y > = 0 \,, \qquad X,Y,Z \in \mathfrak{m}, \tag{6.3}$$

where $<,>$ denotes the inner product of $\mathfrak{m}$ induced on $\mathfrak{m}$ from g by identification of $\mathfrak{m}$ with T_pM. Further $[,]_{\mathfrak{m}}$ is the projection of $[,]$ on $\mathfrak{m}$. (See [26, vol. II, p. 202].) Note that, when K is connected, (6.2) is equivalent to

$$[\mathfrak{k}, \mathfrak{m}] \subseteq \mathfrak{m} \tag{6.4}$$

([26, vol. II, p. 190]) and that K is always connected if M is simply connected.

Following [12] the Riemannian metric g of M is said to be a *naturally reductive metric* on M if M is homogeneous and if there exists a transitive (and effective) group G of isometries of M with the properties (6.1), (6.2), (6.3).

In what follows (M,g) will said to be a *naturally reductive homogeneous Riemannian manifold* if g is a naturally reductive metric. It is clear that it is not always easy to decide whether a homogeneous Riemannian manifold is or is not naturally reductive since one has to consider *all* transitive groups of isometries. For this reason we will now give another characterization using the Ambrose-Singer theorem. In many cases it is much easier to use this result to obtain a quick answer to the problem.

First we recall that, if (M,g) is naturally reductive, the canonical connection $\tilde{\nabla}$ is a metric connection with parallel curvature tensor and parallel torsion tensor (parallel with respect to $\tilde{\nabla}$). Hence $T = \nabla - \tilde{\nabla}$ determines a homogeneous structure on (M,g). Further we have

$$(T_p)_X Y = \frac{1}{2}[X,Y]_{\mathfrak{m}}, \qquad X,Y \in \mathfrak{m} \tag{6.5}$$

([26, vol. II, p. 201]) and this implies that $T \in \mathcal{T}_3$. Hence : *all naturally reductive homogeneous Riemannian manifolds admit a homogeneous structure of type* $\mathcal{T}_3$. Now we prove the converse.

<u>THEOREM 6.1.</u> *Let* (M,g) *be a connected, complete and simply connected Riemannian manifold equipped with a homogeneous structure* T *of type* $\mathcal{T}_3$. *Then* (M,g) *is a naturally reductive homogeneous Riemannian manifold.*

<u>Proof.</u> Let G be the transitive and effective group of isometries of (M,g) obtained from T by using the method given in chapter 1, section C. Let $\mathfrak{g} = \mathfrak{m} \oplus \mathfrak{k}$ be the decomposition (1.76) of the Lie algebra $\mathfrak{g}$ of G. It follows from (1.79) that (6.4) is satisfied. But since M is simply connected, the isotropy subgroup $\mathfrak{k}$ is connected and hence we have (6.2). This proves that M = G/K is a *reductive homogeneous manifold* with *reductive decomposition* $\mathfrak{g} = \mathfrak{m} \oplus \mathfrak{k}$ (i.e. (6.1) and (6.2) hold [26, vol. II, p. 190]).

Hence we only need a proof for (6.3). Let U be the symmetric bilinear form on $\mathfrak{m}$ defined by

$$2 < U(Y,Z),X > = < [X,Y]_{\mathfrak{m}},Z > + < [X,Z]_{\mathfrak{m}},Y > \tag{6.6}$$

where $X,Y,Z \in m$. This implies that U is identically zero if and only if the metric g is naturally reductive. Now using (1.79) we obtain at once

$$2 \langle U(Y,Z),X \rangle = \langle T_X Y - T_Y X, Z \rangle + \langle T_X Z - T_Z X, Y \rangle .$$

Since T_X is skew-symmetric for all $X \in m$ we get

$$\langle U(Y,Z),X \rangle = \langle T_Y Z + T_Z Y, X \rangle .$$

Hence $U = 0$ if and only if $T \in \mathcal{T}_3$.

We may conclude that to find the naturally reductive homogeneous structures on (M,g) we have to find the solutions of the system (AS) which belong to the class $\mathcal{T}_3$ and this is sometimes much easier.

Finally we note that the naturally reductive homogeneous Riemannian manifolds can also be characterized using a property of the geodesics of (M,g). More specifically, let $M = G/K$ be a reductive representation of (M,g), i.e. we have (6.1) and (6.2). Then (6.3) is satisfied if and only if the geodesics of (M,g) are orbits of one-parameter subgroups of G of type exptX where $X \in m$. This results follows from another well-known result which states that (6.3) is satisfied if and only if the Levi Civita connection and the canonical connection (with respect to the decomposition (6.1)) have the same geodesics (see [26, vol. II, chapter VI]). From these results and from Theorem 6.1 one obtains immediately

<u>THEOREM 6.2.</u> [1] *A necessary and sufficient condition that* $T \in \mathcal{T}_3$ *is that every geodesic in* M *is the orbit of a one-parameter subgroup of isometries of the corresponding group* G *with infinitesimal generator in* m.

This theorem has been obtained in [1]. It is Theorem 5.4 of that paper but it is necessary to add there a condition. (See the example of Kaplan in chapter 9.)

As is well-known all irreducible symmetric spaces are naturally reductive but there are also many other examples. In particular all the isotropy irreducible homogeneous manifolds studied by J. Wolf in [57] belong to this class. See also [12] for examples in the class of the compact Lie groups. Finally all nearly Kähler 3-symmetric spaces are naturally reductive (see chapter 8 and [13]).

In the rest of this section we concentrate on the homogeneous structures $T \in \mathcal{T}_3$ when (M,g) is a three-dimensional connected manifold. In the first place we note that if $T \in \mathcal{T}_3$ does not vanish, it determines a nonzero 3-form on M by

$$T_{XYZ} = g(T_X Y, Z) , \quad X,Y,Z \in \mathfrak{X}(M) ,$$

and since $\tilde{\nabla}T = 0$, $\|T\|^2$ is a nonvanishing constant. Hence the manifold M must be *orientable*. We denote by dV the volume form on M (determined uniquely up to sign by $dV(E_1,E_2,E_3) = \pm 1$ for an arbitrary orthonormal frame field (E_1,E_2,E_3)). Hence we put

$$T = \lambda dV \tag{6.7}$$

where λ is a constant.

Now we first consider the case of a Riemannian manifold (M,g) of constant curvature. In that case the condition $\tilde{\nabla}R = 0$, with $\nabla - \tilde{\nabla} = T$, is always satisfied and hence we have

<u>THEOREM 6.3.</u> *Let* (M,g) *be a connected, orientable, three-dimensional manifold of constant curvature. Then all the nonvanishing homogeneous structures of type* $\mathcal{T}_3$ *are given by* $T = \lambda dV$ *where* λ *is constant and* dV *is the volume form on* M.

We conclude that there are an infinite number of homogeneous structures of type $\mathcal{T}_3$ on a three-dimensional connected and orientable space of constant curvature. Two structures T_1 and T_2 of this class are isomorphic if and only if $T_1 = \pm T_2$ since $\varphi^* T = T$ when φ is an orientation-preserving isometry and $\varphi^* T = -T$ when φ changes the orientation.

Next we consider the three-dimensional manifolds which do not have constant curvature. In this case also the existence of a structure of type $\mathcal{T}_3$ imposes strong conditions on the curvature. These are consequences of the condition $\tilde{\nabla}R = 0$ which is now not automatically satisfied. More precisely we have

<u>THEOREM 6.4.</u> *Let* (M,g) *be a connected three-dimensional orientable*

Riemannian manifold which admits a nonvanishing homogeneous structure T *of type* $\mathcal{T}_3$. *Then we have*

(i) (M,g) *has constant curvature,*

or

(ii) *there exists a local orthonormal frame* (E_1,E_2,E_3) *such that the local curvature 2-form* Ω *with respect to this frame is given by*

$$\Omega_{12} = -\lambda^2\theta_1 \wedge \theta_2 \, , \; \Omega_{13} = -\lambda^2\theta_1 \wedge \theta_3 \, , \; \Omega_{23} = (\lambda^2 + \mu)\theta_2 \wedge \theta_3 \, ,$$

where $(\theta_i, \, i=1,2,3)$ *is the dual frame and* μ *is a constant such that* $\mu + 2\lambda^2 \neq 0$ *with* $T = \lambda dV$.

Proof. Let $T \neq 0 \in \mathcal{T}_3$. Then the first and third condition of the system (AS) are satisfied if and only if

$$T = \lambda dV$$

where λ is a constant. Hence we have only to consider ((AS)(ii)). Now let ρ denote the Ricci tensor of (M,g). Then ((AS)(ii)) is equivalent to

$$\nabla_X\rho_{YZ} = -\rho_{T_XYZ} - \rho_{YT_XZ} \, , \qquad X,Y,Z \in \mathfrak{X}(M) \, , \tag{6.8}$$

since M is three-dimensional.

Next let (E_1,E_2,E_3) be a local orthonormal frame field which diagonalizes the Ricci tensor and such that $dV(E_1,E_2,E_3) = 1$. Further put $\rho_{ij} = \rho(E_i,E_j)$. Then (6.8) is equivalent to

$$E_i(\rho_{jk}) = (\rho_{kk} - \rho_{jj})\{g(\nabla_{E_i}E_j,E_k) - \lambda dV(E_i,E_j,E_k)\} \, . \tag{6.9}$$

These conditions imply at once that the eigenvalues of ρ are constant on M and so (6.9) becomes

$$(\rho_{kk} - \rho_{jj})\{g(\nabla_{E_i}E_j,E_k) - \lambda dV(E_i,E_j,E_k)\} = 0. \tag{6.10}$$

This implies that we have to consider three cases :

a) $\rho_{11} = \rho_{22} = \rho_{33}$;

b) $(\rho_{11} - \rho_{22})(\rho_{22} - \rho_{33})(\rho_{33} - \rho_{11}) \neq 0$;

c) $\rho_{11} \neq \rho_{22} = \rho_{33}$.

For a) M is an Einstein manifold and hence of constant curvature.

Next we consider the case b). Here (6.10) becomes

$$g(\nabla_{E_i} E_j, E_k) = \lambda dV(E_i, E_j, E_k). \tag{6.11}$$

This implies that, if (i,j,k) is an even permutation of (1,2,3), then the connection forms of the Levi Civita connection ∇ are given by

$$\omega_{ij} = \lambda\theta_k. \tag{6.12}$$

This is an easy consequence of (6.11) and

$$\omega_{ij}(X) = g(\nabla_X E_i, E_j) = \lambda dV(X, E_i, E_j).$$

Further, using the structure equations of Cartan for this case, we obtain

$$d\theta_i = \omega_{ij} \wedge \theta_j + \omega_{ik} \wedge \theta_k \tag{6.13}$$

and with (6.12) this implies

$$d\theta_i = - 2\lambda\theta_j \wedge \theta_k. \tag{6.14}$$

Next, since

$$\Omega_{ij} = d\omega_{ij} - \omega_{ik} \wedge \omega_{kj} , \tag{6.15}$$

we obtain from (6.12), (6.14) and (6.15) :

$$\Omega_{ij} = - \lambda^2\theta_i \wedge \theta_j. \tag{6.16}$$

Hence (M,g) must be a space of constant curvature, but this is impossible because of condition b).

Finally we examine the case c). First (6.10) implies

$$g(\nabla_{E_i} E_1, E_2) = \lambda dV(E_i, E_1, E_2) ,$$

$$g(\nabla_{E_i} E_1, E_3) = \lambda dV(E_i, E_1, E_3) , \qquad i = 1,2,3,$$

and so we get

$$\omega_{12} = \lambda\theta_3 , \quad \omega_{13} = - \lambda\theta_2. \tag{6.17}$$

Using (6.13) we obtain from (6.17)

$$\begin{cases} d\theta_1 = - 2\lambda\theta_2 \wedge \theta_3 , \\ d\theta_2 = \lambda\theta_1 \wedge \theta_3 + \omega_{23} \wedge \theta_3 , \\ d\theta_3 = -\lambda\theta_1 \wedge \theta_2 - \omega_{23} \wedge \theta_2 , \end{cases} \tag{6.18}$$

and with (6.15) and (6.18) we obtain

$$\begin{cases} \Omega_{12} = - \lambda^2\theta_1 \wedge \theta_2 , \\ \Omega_{13} = - \lambda^2\theta_1 \wedge \theta_3 , \\ \Omega_{23} = d\omega_{23} + \lambda^2\theta_2 \wedge \theta_3. \end{cases} \tag{6.19}$$

Next recall that

$$\Omega_{ij} = - \frac{1}{2} \sum_{p,q} R_{ijpq}\theta_p \wedge \theta_q \tag{6.20}$$

and hence

$$\rho_{ij} = - 2 \sum_m \Omega_{im}(E_j, E_m) . \tag{6.21}$$

Using (6.19) we obtain at once

$$\rho_{11} = 2\lambda^2 ,$$

$$\rho_{22} = \rho_{33} = - 2d\omega_{23}(E_2,E_3) ,$$

$$\rho_{12} = - 2d\omega_{23}(E_1,E_3) ,$$

$$\rho_{13} = - 2d\omega_{23}(E_1,E_2).$$

Since $\rho_{12} = \rho_{13} = 0$ we must have

$$d\omega_{23} = \mu\theta_2 \wedge \theta_3 \tag{6.22}$$

and because $\rho_{22} = \rho_{33} = - \mu$, μ must be constant. The required result now follows from (6.18), (6.19) and (6.22).

Theorem 6.3, Theorem 6.4 and the theorem of Ambrose-Singer make it possible to give the complete classification for dimension three.

<u>THEOREM 6.5.</u> *Let* (M,g) *be a three-dimensional connected complete and simply connected Riemannian manifold which admits a nonvanishing homogeneous structure of type* $\mathcal{T}_3$. *Then* (M,g) *is*

i) $\mathbb{R}^3, S^3$ or $\mathbb{H}^3$;

or

ii) *isometric to one of the following Lie groups with a suitable left invariant metric :*

1. SU(2) ;
2. $\widetilde{SL(2,\mathbb{R})}$, *the universal covering of* $SL(2,\mathbb{R})$;
3. *the Heisenberg group.*

<u>Proof.</u> The result of Ambrose and Singer implies that (M,g) is homogeneous and the Lie algebra of the group G which acts transitively and effectively on M is isomorphic to the direct sum $m \oplus k$ where k is the holonomy algebra of the connection $\tilde{\nabla} = \nabla - T$ and m the tangent space at M in p. This direct sum is equipped with the structure (1.79).

i) is proved in Theorem 6.3. So we consider now case ii).

Let $(E_i, i=1,2,3)$ be a local orthonormal frame in the neighbourhood of p as in Theorem 6.4 and recall that the connection forms of $\tilde{\nabla}$ are related to those of ∇ by

$$\tilde{\omega}_{ij}(X) = g(\tilde{\nabla}_X E_i, E_j) = \omega_{ij}(X) - \lambda dV(X, E_i, E_j). \tag{6.23}$$

Then (6.17) implies

$$\tilde{\omega}_{12} = 0 \ , \quad \tilde{\omega}_{13} = 0 \ , \quad \tilde{\omega}_{23} = \omega_{23} - \lambda\theta_1. \tag{6.24}$$

From the structure equations we derive

$$\tilde{\Omega}_{12} = 0 \ , \quad \tilde{\Omega}_{13} = 0 \ , \quad \tilde{\Omega}_{23} = (2\lambda^2 + \mu)\theta_2 \wedge \theta_3 \tag{6.25}$$

and using

$$\tilde{\Omega}_{ij} = -\frac{1}{2} \sum_{p,q} \tilde{R}_{pqij} \theta_p \wedge \theta_q \ , \tag{6.26}$$

we obtain

$$\tilde{R} = -4(2\lambda^2 + \mu)(\theta_2 \wedge \theta_3) \otimes (\theta_2 \wedge \theta_3). \tag{6.27}$$

Hence we have

$$\begin{cases} (\tilde{R}_{E_1E_2})_p = (\tilde{R}_{E_1E_3})_p = 0 \ , \\ (\tilde{R}_{E_2E_3})_p = (2\lambda^2 + \mu)\begin{pmatrix} 0 & 0 & 0 \\ 0 & 0 & 1 \\ 0 & -1 & 0 \end{pmatrix} = (2\lambda^2 + \mu)\phi. \end{cases} \tag{6.28}$$

Next let $e_i = (E_i)_p$, $i=1,2,3$. Then (e_1, e_2, e_3, ϕ) is a basis for $\mathfrak{g}$ and $\mathfrak{k}$ is generated by ϕ. Further (1.79) implies

$$\begin{cases} [e_1, e_2] = 2\lambda e_3 \ , & [e_1, \phi] = 0 \ , \\ [e_1, e_3] = -2\lambda e_2 \ , & [e_2, \phi] = e_3 \ , \end{cases} \tag{6.29}$$

$$\left| \; [e_2,e_3] = 2\lambda e_1 + (2\lambda^2 + \mu)\phi, \quad [e_3,\phi] = - e_2. \right.$$

Since $T \neq 0$ we have $\lambda \neq 0$. So we can make the following transformation :

$$\begin{cases} u_1 = e_1 + \dfrac{2\lambda^2 + \mu}{2\lambda}\phi , \\ u_2 = e_2 , \\ u_3 = e_3 . \end{cases} \tag{6.30}$$

Then g is generated by u_1,u_2,u_3,ϕ and we have

$$\begin{cases} [u_2,u_3] = 2\lambda u_1 , \\ [u_3,u_1] = \dfrac{2\lambda^2 - \mu}{2\lambda} u_2 , \\ [u_1,u_2] = \dfrac{2\lambda^2 - \mu}{2\lambda} u_3 ; \end{cases} \tag{6.31}$$

$$\begin{cases} [u_1,\phi] = 0 , \\ [u_2,\phi] = u_3 , \\ [u_3,\phi] = - u_2 . \end{cases} \tag{6.32}$$

Let h denote the subspace of g generated by u_1,u_2,u_3. Then (6.31) implies that h is a 3-dimensional *unimodular* subalgebra of g (see [33, p. 305]). Moreover, (6.32) shows that g is a semi-direct product of h and k.

Let H denote the connected and simply connected group corresponding to h. We prove that M is isometric to H where H is equipped with the left invariant metric g' such that (u_1,u_2,u_3) is an orthonormal basis. Here u_1,u_2,u_3 denote the left invariant vector fields which correspond to the basis (u_1,u_2,u_3) at p.

Let $\hat{\nabla}$ denote the Riemannian connection of H with respect to the metric g'. Then we have

$$2g'(\hat{\nabla}_X Y,Z) = g'([X,Y],Z) - g'([Y,Z],X) + g'([Z,X],Y)$$

where X,Y and Z are left invariant vector fields on H (see [33]). This implies for the connection forms of $\nabla' = \hat{\nabla} - T$ with $T = \lambda dV$:

$$\omega'_{12} = 0 \ , \quad \omega'_{13} = 0 \ , \quad \omega'_{23} = -\frac{2\lambda^2 + \mu}{2\lambda}\psi_1 \tag{6.33}$$

where (ψ_1,ψ_2,ψ_3) are the dual left invariant 1-forms of (u_1,u_2,u_3). Then (6.31) implies

$$\begin{cases} d\psi_1 = -2\lambda\,\psi_2 \wedge \psi_3 \ , \\ d\psi_2 = \dfrac{2\lambda^2 - \mu}{2\lambda}\psi_1 \wedge \psi_3 \ , \\ d\psi_3 = -\dfrac{2\lambda^2 - \mu}{2\lambda}\psi_1 \wedge \psi_2 \ , \end{cases} \tag{6.34}$$

and so we obtain from the structure equations :

$$\Omega'_{12} = \Omega'_{13} = 0 \ , \quad \Omega'_{23} = (2\lambda^2 + \mu)\psi_2 \wedge \psi_3. \tag{6.35}$$

Hence

$$R' = -4(2\lambda^2 + \mu)(\psi_2 \wedge \psi_3) \otimes (\psi_2 \wedge \psi_3) \ . \tag{6.36}$$

To compute the torsion tensor S', written as a (0,3)-tensor, we use

$$S' = 2\sum_m \Theta_m \otimes \psi_m \tag{6.37}$$

where the torsion forms Θ_m are defined by

$$\Theta_k = d\psi_k - \sum_m \omega'_{km} \wedge \psi_m. \tag{6.38}$$

This gives :

$$S' = -12\lambda\,\psi_1 \wedge \psi_2 \wedge \psi_3 \ . \tag{6.39}$$

Further we have

$$\begin{cases} \nabla'_X\psi_1 = 0 , \\ \nabla'_X\psi_2 = -\dfrac{2\lambda^2 + \mu}{2\lambda}\psi_1(X)\psi_3 , \\ \nabla'_X\psi_3 = \dfrac{2\lambda^2 + \mu}{2\lambda}\psi_1(X)\psi_2 \end{cases} \tag{6.40}$$

where $X \in \mathfrak{X}(H)$, and this implies with (6.36) and (6.39) that $\nabla'R' = \nabla'S' = 0$.

Finally, let $f : T_pM \to T_eH = h$ be the linear map defined by

$$f(e_i) = u_i , \qquad i = 1,2,3. \tag{6.41}$$

f is an isometry and $f^*(\psi_i)_p - (\theta_i)_p = 0$, $i = 1,2,3$. Then (6.27) and (6.36) imply

$$f^*(R'_p) = \tilde{R}_p .$$

Further, the torsion of $\tilde{\nabla}$ is given by $\tilde{S} = - 2T = - 2\lambda dV$ and so we have

$$f^*(S'_p) = \tilde{S}_p$$

since $dV = 3!\theta_1 \wedge \theta_2 \wedge \theta_3$. Hence we conclude that f is a linear map which preserves the curvature and the torsion tensor of $\tilde{\nabla}$ and of ∇'. These connections are complete since both are metric and both manifolds (M,g) and (H,g') are complete. So, using a standard argument (see [26, vol. I, p. 265]) we know that there exists an affine transformation $\varphi : M \to H$ such that $\varphi(p) = e$ and $(\varphi_*)_p = f$. Hence, since $\tilde{\nabla}$ and ∇' are metric and $(\varphi_*)_p = f$ is an isometry, the map φ is an isometry (see the note after Proposition 1.3).

We are left with the classification of the groups H. To do this we use the results of Milnor [33]. There are three cases to consider :

a) $\mu = 2\lambda^2$;
b) $\lambda(2\lambda^2 - \mu) > 0$;
c) $\lambda(2\lambda^2 - \mu) < 0$.

a) corresponds to the Heisenberg group, b) to SU(2) and c) to $\widetilde{SL(2,\mathbb{R})}$. This finishes the proof.

In the last part of this chapter we prove a theorem concerning the existence of homogeneous structures of type $\mathcal{T}_1 \oplus \mathcal{T}_3$ on three-dimensional Riemannian manifolds. This theorem implies that to obtain an example of type $\mathcal{T}_1 \oplus \mathcal{T}_3$, which is not of type $\mathcal{T}_1$ or $\mathcal{T}_3$, we have at least to consider four-dimensional manifolds (see chapter 8).

THEOREM 6.6. *Let* (M,g) *be a connected three-dimensional manifold which admits a homogeneous structure* T *of type* $\mathcal{T}_1 \oplus \mathcal{T}_3$. *Then* $T \in \mathcal{T}_1$ *or* $T \in \mathcal{T}_3$.

Proof. Since the proof is local in nature and since M is locally orientable we can write

$$T_{XYZ} = g(X,Y)g(\xi,Z) - g(Y,\xi)g(X,Z) + \lambda dV(X,Y,Z) \tag{6.42}$$

for $X,Y,Z \in \mathfrak{X}(M)$. Further $\tilde{\nabla}T = 0$ is equivalent to the conditions $\tilde{\nabla}\xi = 0$ and λ = constant. Hence $g(\xi,\xi)$ = constant.

Next using

$$g(X \wedge Y,Z) = dV(X,Y,Z) \;, \tag{6.43}$$

$\tilde{\nabla}\xi = 0$ can be rewritten as

$$\nabla_X\xi = \lambda X \wedge \xi + g(X,\xi)\xi - g(\xi,\xi)X. \tag{6.44}$$

Since

$$\nabla_X(Y \wedge Z) = \nabla_X Y \wedge Z + Y \wedge \nabla_X Z \tag{6.45}$$

and

$$\mathfrak{S}_{X,Y,Z}\; X \wedge Y \wedge Z = 0 \;, \tag{6.46}$$

we obtain from (6.44) :

$$R_{XYZ\xi} = 2\lambda dV(X,Y,\xi)g(Z,\xi) + 2\lambda^2 dV(X \wedge Y,\xi,Z) \qquad (6.47)$$

$$- 2\lambda dV(X,Y,Z)g(\xi,\xi)$$

$$- g(\xi,\xi)\{g(X,\xi)g(Y,Z) - g(Y,\xi)g(X,Z)\}.$$

Using the first Bianchi identity $\underset{X,Y,Z}{\mathfrak{S}} R_{XYZ\xi} = 0$ we derive from (6.47) :

$$\lambda\{ \underset{X,Y,Z}{\mathfrak{S}} dV(X,Y,\xi)g(Z,\xi) - 3g(\xi,\xi)dV(X,Y,Z)\} = 0 . \qquad (6.48)$$

$\lambda = 0$ implies $T \in \mathcal{T}_1$. Hence we may suppose $\lambda \neq 0$. Then we take an orthonormal frame (E_1,E_2,E_3) such that $dV(E_1,E_2,E_3) = 1$. Further let (i,j,k) be an even permutation of $(1,2,3)$. For $\xi = \sum \xi_i E_i$ we have

$$dV(E_i,E_j,\xi)g(\xi,E_k) = \xi_k^2 . \qquad (6.49)$$

Finally, using (6.49) we derive from (6.48) that $g(\xi,\xi) = 0$. This means that $T \in \mathcal{T}_3$.

REMARKS

A. Among the homogeneous manifolds the naturally reductive spaces are the most simple ones. In Theorem 6.1 we have shown how these spaces can be characterized using the tensors T.

In what follows we give an application of this theorem. We will show how naturally reductive homogeneous spaces can be characterized by means of the *volume of geodesic spheres or geodesic balls*. This property is an easy consequence of the theory developed in [34],[35] which is a generalization of the theory given in [16].

Let (M,g) be a Riemannian manifold of dimension n and denote by $B_m^D(r)$ the geodesic ball with center m and sufficiently small radius r, where the geodesics are those of a metric connection D. In [16] a power series expansion is given for the volume $V_m^D(r)$ of $B_m^D(r)$ when D is the Levi Civita connection ∇ and this formula is generalized in [34],[35] when D is a arbitrary metric connection. If

$$D - \nabla = T$$

and DT = 0, then we have explicitly the following expansion :

$$V_m^D(r) = \omega r^n\{1 + Ar^2 + 0(r^4)\}$$

where $\omega = \dfrac{(\pi r^2)^{n/2}}{(\frac{n}{2})!}$ is the volume of a unit ball in Euclidean space of dimension n and

$$A = \frac{-1}{6(n+2)}\left\{\tau + \frac{1}{4}(\|T\|^2 + < T,\hat{T} > + \|c_{12}(T)\|^2)\right\}_m.$$

Here τ denotes the scalar curvature of the Levi Civita connection. The following theorem now follows at once using Theorem 3.3 and Theorem 3.5.

THEOREM 6.7. *Let* (M,g) *be a simply-connected, connected and complete Riemannian manifold. Then* (M,g) *is a naturally reductive homogeneous manifold if and only if there exists a tensor* T *satisfying*

i) $\tilde{\nabla}R = 0$,

ii) $\tilde{\nabla}T = 0$,

where $\tilde{\nabla} = \nabla - T$ *is a metric connection, and*

iii) $V_m^{\tilde{\nabla}}(r) = V_m^{\nabla}(r)$ *for all* $m \in M$ *and all small* r.

Note that the geodesics of $\tilde{\nabla}$ and ∇ on the homogeneous space are the same if and only if M is naturally reductive. In that case the geodesic spheres coincide and hence have the same volume.

Further we have, when DT = 0 :

$$\tau_D = \tau + \frac{1}{2} < T,\hat{T} > + \|c_{12}(T)\|^2$$

where τ_D denotes the scalar curvature of the connection D. It follows easily that among the naturally reductive homogeneous spaces, the symmetric spaces are characterized by the condition $\tau_{\tilde{\nabla}} = \tau$.

B. Let M be a differentiable manifold and denote by $\tilde{\nabla}$ and ∇ two linear connections on M. $\tilde{\nabla}$ and ∇ are said to be *projectively equivalent* if and only if they have the same paths (or autoparallel curves). From this it is clear that Theorem 6.1 and the remark above this theorem give the following characterization for naturally reductive homogeneous spaces.

THEOREM 6.8. *Let* (M,g) *be a connected, complete and simply connected Riemannian manifold. Then* (M,g) *is a naturally reductive homogeneous manifold if and only if there exists a tensor field* T *of type* (1,2) *satisfying the conditions* (AS) *and such that* $\tilde{\nabla}$ *and* ∇ *are projectively equivalent.*

7. THE HEISENBERG GROUP

The *Heisenberg group* H is the subgroup of GL(3, $\mathbb{R}$) formed by the matrices of type

$$\begin{pmatrix} 1 & x & y \\ 0 & 1 & z \\ 0 & 0 & 1 \end{pmatrix}.$$

This Lie group is diffeomorphic to $\mathbb{R}^3$. It is easy to see that

$$g = dx^2 + dz^2 + (dy - xdz)^2 \qquad (7.1)$$

is a left invariant metric on H. In this chapter we give a detailed study of this homogeneous space (H,g). Our main purpose is to provide a nice example which illustrates the results and methods given in earlier chapters of this paper.

More specifically we shall prove that (H,g) admits an infinite number of homogeneous structures T(μ), $\mu \in \mathbb{R}$. All these structures are nonisomorphic. For $\mu = -\frac{1}{2}$ the corresponding group of isometries of (H,g) is the group H itself, acting on the left on H. In all the other cases the corresponding group G is the connected component of the identity of the full group of isometries of (H,g) and this is a semi-direct product of H with K = SO(2). The existence of the different nonisomorphic homogeneous structures associated with the same group G comes from the fact that there are an infinite number of reductive decompositions of the Lie algebra $\mathfrak{g} = \mathfrak{m} \oplus \mathfrak{k}$ of G.

Also we shall prove that all the homogeneous structures belong to the class $\mathcal{T}_2 \oplus \mathcal{T}_3$. Further T($\mu$) $\in \mathcal{T}_2$ if and only if $\mu = -1$

and $T(\mu) \in \mathcal{T}_3$ if and only if $\mu = \frac{1}{2}$. This proves that (H,g) is a naturally reductive homogeneous space. Further it shows that the following inclusions

$$\begin{array}{ccccc} & & \mathcal{T}_2 & & \\ & \subset & & \subset & \\ \{0\} & & & & \mathcal{T}_2 \oplus \mathcal{T}_3 \\ & \subset & & \subset & \\ & & \mathcal{T}_3 & & \end{array}$$

are strict inclusions.

First we start with the calculation of the Levi Civita connection and the curvature of (H,g). We put

$$\theta_1 = dx\ , \quad \theta_2 = dz\ , \quad \theta_3 = dy - xdz. \tag{7.2}$$

The dual basis is given by

$$E_1 = \frac{\partial}{\partial x}\ , \quad E_2 = \frac{\partial}{\partial z} + x\,\frac{\partial}{\partial y}\ , \quad E_3 = \frac{\partial}{\partial y}\ . \tag{7.3}$$

Note that the 1-forms θ_i, and hence the vector fields E_i, $i=1,2,3$, are left invariant. So (E_1,E_2,E_3) is a basis for the Lie algebra of H. Further the Levi Civita connection is determined by

$$g(\nabla_X E_i, E_j) = \omega_{ij}(X) \tag{7.4}$$

where the connection forms are uniquely determined by the structure equations of Cartan. From (7.2) we obtain at once

$$d\theta_1 = d\theta_2 = 0\ , \quad d\theta_3 = -\,\theta_1 \wedge \theta_2 \tag{7.5}$$

and this implies

$$\omega_{12} = -\frac{1}{2}\,\theta_3\ , \quad \omega_{13} = -\frac{1}{2}\,\theta_2\ , \quad \omega_{23} = \frac{1}{2}\,\theta_1. \tag{7.6}$$

From the second group of structure equations we get for the curvature 2-forms :

$$\Omega_{12} = \frac{3}{4}\,\theta_1 \wedge \theta_2\,, \quad \Omega_{13} = -\frac{1}{4}\,\theta_1 \wedge \theta_3\,, \quad \Omega_{23} = -\frac{1}{4}\,\theta_2 \wedge \theta_3. \tag{7.7}$$

Hence

$$R_{1212} = -\frac{3}{4}\,, \quad R_{1313} = R_{2323} = \frac{1}{4}\,, \tag{7.8}$$

and from (7.8) we obtain for the Ricci tensor :

$$\rho_{11} = \rho_{22} = -\frac{1}{2}\,, \quad \rho_{33} = \frac{1}{2}\,, \tag{7.9}$$

the other components being zero.

Next we determine all the homogeneous structures on (H,g). We have

THEOREM 7.1. *Let* (H,g) *denote the Heisenberg group with the left invariant metric* (7.1). *Then all the homogeneous structures* T *on* (H,g) *are given by*

$$T(\mu) = 2\mu\,\theta_3 \otimes (\theta_1 \wedge \theta_2) + \theta_1 \otimes (\theta_2 \wedge \theta_3) + \theta_2 \otimes (\theta_3 \wedge \theta_1) \tag{7.10}$$

where θ_i, i=1,2,3, *are the* 1-*forms given by* (7.2). μ *is an arbitrary constant.*

Proof. Let T be a (0,3)-tensor field such that

$$T_{XYZ} + T_{XZY} = 0$$

for all $X,Y,Z \in \mathfrak{X}(H)$. Denoting the corresponding (1,2)-tensor field by the same symbol T, we put $\nabla - \widetilde{\nabla} = T$.

Since (H,g) is a three-dimensional homogeneous space, the condition $\widetilde{\nabla}R = 0$ is equivalent to $\widetilde{\nabla}\rho = 0$ or equivalently

$$(\nabla_X\rho)_{YZ} = -\rho_{T_XYZ} - \rho_{YT_XZ} \tag{7.11}$$

for $X,Y,Z \in \mathfrak{X}(H)$. Since (E_1,E_2,E_3) diagonalizes ρ, (7.11) is equivalent to

$$(\rho_{jj} - \rho_{ii})(\omega_{ij}(X) - T_{XE_2E_3}) = 0. \tag{7.12}$$

Using (7.6) and (7.9) this becomes

$$T_{XE_1E_3} = \omega_{13}(X) \ , \quad T_{XE_2E_3} = \omega_{23}(X). \tag{7.13}$$

Next we consider the third condition of Ambrose-Singer, namely $\widetilde{\nabla}T = 0$, to determine $\alpha(X) = T_{XE_1E_2}$. Using (7.13) we can write

$$T = 2\alpha \otimes (\theta_1 \wedge \theta_2) + \theta_1 \otimes (\theta_2 \wedge \theta_3) + \theta_2 \otimes (\theta_3 \wedge \theta_1). \tag{7.14}$$

Further, the connection forms of $\widetilde{\nabla}$ are determined by

$$\widetilde{\omega}_{ij}(X) = \omega_{ij}(X) - T_{XE_iE_j} \ ,$$

$X \in \mathfrak{X}(H)$. It follows easily that

$$\widetilde{\omega}_{13} = \widetilde{\omega}_{23} = 0 \ , \quad \widetilde{\omega}_{12} = \omega_{12} - \alpha = -(\alpha + \frac{1}{2}\theta_3). \tag{7.15}$$

Now we have also

$$\begin{cases} \widetilde{\nabla}_X\theta_1 = -(\alpha + \frac{1}{2}\theta_3)(X)\theta_2 \ , \\ \widetilde{\nabla}_X\theta_2 = (\alpha + \frac{1}{2}\theta_3)(X)\theta_1 \ , \\ \widetilde{\nabla}_X\theta_3 = 0. \end{cases} \tag{7.16}$$

Using (7.16) we obtain from (7.14) :

$$\widetilde{\nabla}_X T = 2\widetilde{\nabla}_X\alpha \otimes (\theta_1 \wedge \theta_2) \tag{7.17}$$

and hence $\widetilde{\nabla}_X T = 0$ is equivalent to $\widetilde{\nabla}_X\alpha = 0$. Therefore, with

$$\alpha = \sum_{i=1}^{3} \alpha_i\theta_i \ ,$$

we must have

$$\begin{cases} X(\alpha_1) = -\alpha_2(\alpha + \frac{1}{2}\theta_3)(X) , \\ X(\alpha_2) = \alpha_1(\alpha + \frac{1}{2}\theta_3)(X) , \\ X(\alpha_3) = 0. \end{cases} \tag{7.18}$$

Hence α_3 is a constant. Further, using the integrability conditions

$$(XY - YX - [X,Y])(\alpha_i) = 0 , \qquad i = 1,2,$$

we obtain from (7.18) :

$$\begin{cases} \alpha_1 d(\alpha + \frac{1}{2}\theta_3) = 0 , \\ \alpha_2 d(\alpha + \frac{1}{2}\theta_3) = 0. \end{cases} \tag{7.19}$$

So we have $\alpha_1 = \alpha_2 = 0$ or $d(\alpha + \frac{1}{2}\theta_3) = 0$. In the last case we can write

$$\alpha + \frac{1}{2}\theta_3 = df$$

where f is a C^∞ function. This implies

$$\alpha_1 = E_1(f) , \quad \alpha_2 = E_2(f) , \quad \alpha_3 = E_3(f) - \frac{1}{2} .$$

So $E_3(f)$ is constant. Further (7.18) implies

$$\begin{cases} XE_1(f) = -E_2(f)df(X) , \\ XE_2(f) = E_1(f)df(X) \end{cases} \tag{7.20}$$

and since

$$[E_1,E_2] = E_3 , \quad [E_1,E_3] = [E_2,E_3] = 0 , \tag{7.21}$$

we must have $E_1(f) = E_2(f) = 0$. Putting $\mu = \alpha_3$, we obtain the required result.

<u>THEOREM 7.2.</u> *Let* $T(\mu)$ *be a homogeneous structure on* (H,g). *Then*

$T \in \mathcal{T}_2 \oplus \mathcal{T}_3$. *Further* $T \in \mathcal{T}_2$ *if and only if* $\mu = -1$ *and* $T \in \mathcal{T}_3$ *(or equivalently,* (H,g) *is naturally reductive) if and only if* $\mu = \frac{1}{2}$.

Proof. (7.10) implies

$$T_{E_i} E_i = 0 \ , \quad i = 1,2,3,$$

and

$$\begin{cases} T_{E_1} E_2 = - T_{E_2} E_1 = \frac{1}{2} E_3 \ , \\ T_{E_1} E_3 = - \frac{1}{2} E_2 \ , \quad T_{E_3} E_1 = \mu E_2 \ , \\ T_{E_2} E_3 = \frac{1}{2} E_1 \ , \quad T_{E_3} E_2 = - \mu E_2 . \end{cases} \tag{7.22}$$

The result now follows at once by using the defining conditions given in chapter 3.

In what follows we construct explicitly the Lie algebra $\mathfrak{g}$ of the transitive and effective group G of isometries of (H,g) associated with the homogeneous structure $T(\mu)$. We recall that $\mathfrak{g}$ is isomorphic to the direct sum $m \oplus k$, where k is the holonomy algebra of $\widetilde{\nabla}$ and $m = T_p H$, $p \in H$. In what follows we shall take for p the origin 0 of H. This direct sum has the structure given by (1.79) and k is generated by the skew-symmetric operators $(\widetilde{R}_0)_{XY}$ of m.

Using (7.15) and the structure equations we obtain

$$\widetilde{\Omega}_{13} = \widetilde{\Omega}_{23} = 0 \ , \quad \widetilde{\Omega}_{12} = (\mu + \tfrac{1}{2})\theta_1 \wedge \theta_2 \tag{7.23}$$

where we have put $\alpha = \mu\theta_3$. Hence $\widetilde{R}_{13} = \widetilde{R}_{23} = 0$ and so k is generated by

$$(\widetilde{R}_0)_{12} = (\mu + \tfrac{1}{2}) \begin{pmatrix} 0 & 1 & 0 \\ -1 & 0 & 0 \\ 0 & 0 & 0 \end{pmatrix}. \tag{7.24}$$

Next, put $e_i = (E_i)_0$, i=1,2,3 and $e_4 = (\widetilde{R}_0)_{12}$. Then (1.79) and (7.22) imply

$$\begin{cases} [e_1,e_2] = e_3 + e_4 , & [e_1,e_4] = (\mu + \frac{1}{2})e_2 , \\ [e_1,e_3] = -(\mu + \frac{1}{2})e_2 , & [e_2,e_4] = -(\mu + \frac{1}{2})e_1 , \\ [e_2,e_3] = (\mu + \frac{1}{2})e_1 , & [e_3,e_4] = 0. \end{cases} \tag{7.25}$$

First, let $\mu + \frac{1}{2} = 0$. Then k has dimension zero and g is isomorphic to the Lie algebra h of the Heisenberg group.

In the case $\mu + \frac{1}{2} \neq 0$ we put

$$u_1 = e_1 , \quad u_2 = e_2 , \quad u_3 = e_3 + e_4 , \quad u_4 = (\mu + \tfrac{1}{2})^{-1} e_4.$$

Then (7.25) becomes

$$\begin{cases} [u_1,u_2] = u_3, & [u_1,u_4] = u_2 , \\ [u_1,u_3] = 0 , & [u_2,u_4] = -u_1 , \\ [u_2,u_3] = 0 , & [u_3,u_4] = 0. \end{cases} \tag{7.26}$$

Hence g is a four-dimensional Lie algebra. $\mathcal{L}(u_1,u_2,u_3)$ is a subalgebra which is isomorphic to the Lie algebra h of the Heisenberg group. Further it follows from (7.26) that g is a semi-direct sum of $\mathcal{L}(u_1,u_2,u_3)$ and k.

We conclude that, when $\mu = -\frac{1}{2}$, the corresponding group of isometries G is the Heisenberg group itself. In the other case G is a semi-direct product of H with a one-dimensional Lie group. More precisely, G is the connected component $\mathfrak{J}_0(H)$ of the full group of isometries $\mathfrak{J}(H)$ of H.

To describe this $\mathfrak{J}_0(H)$ we identify H with $\mathbb{C} \times \mathbb{R}$ by $(x,y,z) \mapsto (w,t)$ where

$$w = x + iz , \quad t = y - \frac{1}{2} xz , \tag{7.27}$$

and $\mathbb{C} \times \mathbb{R}$ is equipped with the product

$$(w,t)(w',t') = (w + w', t + t' + \frac{1}{2}\,\mathrm{Im}(\bar{w}w')). \tag{7.28}$$

Note that (x,z,t) is the coordinate system of H induced by the exponential map since

$$\begin{pmatrix} 1 & x & y \\ 0 & 1 & z \\ 0 & 0 & 1 \end{pmatrix} = \exp\begin{pmatrix} 0 & x & t \\ 0 & 0 & z \\ 0 & 0 & 0 \end{pmatrix}.$$

It follows (see for example [22],[54]) that $\mathfrak{I}_0(H)$ is the semi-direct product $H \times_\varphi SO(2)$ where the action φ of $SO(2)$ on H is given by

$$\varphi(e^{i\theta})(w,t) = (e^{-i\theta}w, t). \tag{7.30}$$

This proves

<u>THEOREM 7.3</u>. *Let* (H,g) *be the Heisenberg group* H *with the homogeneous structure* $T(\mu)$. *For* $\mu = -\frac{1}{2}$ *the corresponding group of isometries is* H *itself and for all the other values of* μ *it is* $H \times_\varphi SO(2)$ *where* φ *is given by* (7.30).

Now we study the behaviour of the homogeneous structures under the action of the isometries of (H,g). We have

<u>THEOREM 7.4</u>. *All the homogeneous structures of* (H,g) *are invariant under the action of* $\mathfrak{I}(H)$.

<u>Proof</u>. The elements of $\mathfrak{I}(H)$ are the left translations L_a, $a \in H$, the isometries $\varphi(e^{i\theta})$ given by (7.30), the transformation ψ defined by

$$\psi(w,t) = (\bar{w}, -t) \tag{7.31}$$

and their compositions.

Since θ_i, $i=1,2,3$, are left invariant we have from (7.10) that $L_a^*T(\mu) = T(\mu)$, $a \in H$. Further (7.27) implies

$$\begin{cases} 2\theta_1 = dw + d\bar{w} \, , \\ 2i\theta_2 = dw - d\bar{w} \, , \\ 4i\theta_3 = 4idt + wd\bar{w} - \bar{w}dw \, . \end{cases} \tag{7.32}$$

Hence $\psi^*\theta_1 = \theta_1$, $\psi^*\theta_2 = -\theta_2$ and $\psi^*\theta_3 = -\theta_3$. This implies $\psi^*T(\mu) = T(\mu)$. Finally, we write $T(\mu)$ as follows :

$$2iT(\mu) = \mu\theta_3 \otimes (d\bar{w} \wedge dw) + d\bar{w} \otimes (dw \wedge \theta_3) - dw \otimes (d\bar{w} \wedge \theta_3).$$

Then (7.30) implies $\varphi(e^{i\theta})^*T(\mu) = T(\mu)$ since $\varphi(e^{i\theta})^*\theta_3 = \theta_3$. This proves the required result.

Finally we explain the existence of an infinite number of nonisomorphic homogeneous structures associated with the same group $G = \mathfrak{J}_0(H)$. We show that this is due to the fact that there exist an infinite number of decompositions $\mathfrak{g} = \mathfrak{m} \oplus \mathfrak{k}$ of the Lie algebra $\mathfrak{g}$ of G such that $[\mathfrak{k},\mathfrak{m}] \subseteq \mathfrak{m}$, where $\mathfrak{k}$ is the Lie algebra of SO(2).

Let (u_1,u_2,u_3,u_4) be a basis of $\mathfrak{g}$ such that $\mathfrak{k}$ is generated by u_4 and where the brackets are given by (7.26). The complements of $\mathfrak{k}$ in $\mathfrak{g}$ are the kernels of a nonvanishing linear form f. Since $\mathfrak{k}$ is one-dimensional $[\mathfrak{k},\mathfrak{m}] \subseteq \mathfrak{m}$ implies $f([k,\mathfrak{g}]) = 0$. Hence it follows from (7.26) that we must have

$$f = au^3 + bu^4$$

where $(u^i, i=1,\dots,4)$ denotes the dual basis of (u_i). Since ker f is generated by $u_1,u_2,bu_3 - au_4$, ker f is a complement if and only if $b \neq 0$. Hence all the possible subspaces $\mathfrak{m}$ of $\mathfrak{g}$ are the spaces $\mathfrak{m}_\lambda$ determined by $(u_1,u_2,u_3 + \lambda u_4)$, where λ is an arbitrary constant. Each subspace $\mathfrak{m}_\lambda$ is identified with the tangent space T_0H by the isomorphism φ such that

$$\varphi(u_1) = (E_1)_0 \, , \quad \varphi(u_2) = (E_2)_0 \, , \quad \varphi(u_3 + \lambda u_4) = (E_3)_0 .$$

So the metric g on H induces an inner product $\langle \, , \rangle_\lambda$ on $\mathfrak{m}_\lambda$ such that $(u_1,u_2,u_3 + \lambda u_4)$ is an orthonormal basis.

Next let ∇_λ be the canonical connection of the homogeneous space $H \times_\varphi SO(2)/SO(2)$ associated with the decomposition $\mathfrak{g} = \mathfrak{m}_\lambda \oplus \mathfrak{k}$. Then the tensor $T = \nabla - \nabla_\lambda$ determines a homogeneous structure on (H,g) (see [26, vol. II, p. 193]). A simple calculation shows that

$$T_\lambda = -(1 + 2\lambda)u^3 \otimes (u^1 \wedge u^2) + u^2 \otimes (u^3 \wedge u^1) + u^1 \otimes (u^2 \wedge u^3).$$

Hence the identification $\mathfrak{m}_\lambda = T_0H$ gives all the homogeneous structures of Theorem 7.1.

Note also that if $\lambda_1 \neq \lambda_2$, there are no Lie algebra isomorphisms $\psi : \mathfrak{g} \to \mathfrak{g}$ such that $\psi(\mathfrak{k}) = \mathfrak{k}$, $\psi(\mathfrak{m}_{\lambda_1}) = \mathfrak{m}_{\lambda_2}$ and where $\psi|_{\mathfrak{m}_{\lambda_1}} : \mathfrak{m}_{\lambda_1} \to \mathfrak{m}_{\lambda_2}$ is an isometry. Indeed, an easy calculation shows that all the isomorphisms such that $\psi(\mathfrak{k}) = \mathfrak{k}$ are of the form

$$\begin{pmatrix} A_{11} & A_{12} & 0 & 0 \\ -aA_{12} & aA_{11} & 0 & 0 \\ 0 & 0 & A_{33} & 0 \\ 0 & 0 & 0 & a \end{pmatrix}$$

with respect to the basis (u_1,u_2,u_3,u_4), where

$$a^2 = 1 , \quad A_{33} = a(A_{11}^2 + A_{12}^2) \neq 0 .$$

Then $\psi(\mathfrak{m}_{\lambda_1}) = \mathfrak{m}_{\lambda_2}$ if and only if $\psi(u_3 + \lambda_1 u_4) = c(u_3 + \lambda_2 u_4)$. This implies

$$a\lambda_1 = A_{33}\lambda_2 .$$

Further, since $\psi|_{\mathfrak{m}_{\lambda_1}}$ must be an isometry, we must have

$$\|\psi(u_1)\|_{\lambda_2} = \|u_1\|_{\lambda_1}$$

and this implies

$$A_{11}^2 + a^2 A_{12}^2 = 1.$$

But, since $a^2 = 1$, we get $A_{33} = a$ and so $\lambda_1 = \lambda_2$. This agrees with the results of chapter 2.

8. EXAMPLES AND THE INCLUSION RELATIONS

The main purpose of this chapter is to determine some examples of homogeneous structures. We use these examples to show that the inclusion relations for the eight classes are all strict. We do not go into too much detail although such a detailed study is very instructive. Here we treat the following cases : the Lie groups of dimension three, the k-symmetric spaces (in particular for k=3), the four-dimensional hyperbolic space and some four-dimensional Lie groups.

A. HOMOGENEOUS STRUCTURES ON THREE-DIMENSIONAL LIE GROUPS

Let G be a Lie group equipped with a left invariant metric. Further, let X,Y and Z be left invariant vector fields and define a tensor field T of type (1,2) by

$$2g(T_XY,Z) = g([X,Y],Z) - g([Y,Z],X) + g([Z,X],Y). \tag{8.1}$$

Next consider the metric connection $\tilde{\nabla} = \nabla - T$. Then $\tilde{\nabla}_XY = 0$ for left invariant vector fields X,Y. Hence $\tilde{R} = 0$ and so it is easy to see that T determines a homogeneous structure on (G,g) (see [33]). Note that this connection $\tilde{\nabla}$ is the (-)-connection of Cartan-Schouten.

In what follows we shall only consider this homogeneous structure and we restrict to the case $T \neq 0$.

First we consider the case of a *unimodular* Lie group G. It is well-known that a Lie group is unimodular if and only if the linear transformation Adg has determinant ± 1 for every g in G. If dim G = 3 it is shown in [33] that the Lie algebra $\mathfrak{g}$ of G has an orthonormal basis (e_1,e_2,e_3) such that

$$[e_1,e_2] = \lambda_3e_3\ , \quad [e_2,e_3] = \lambda_1e_1\ , \quad [e_3,e_1] = \lambda_2e_2. \tag{8.2}$$

Hence, using (8.1) and (8.2), we find immediately that

$$g(T_{e_i}e_i,z) = g([e_i,z],e_i) = 0 , \quad i = 1,2,3,$$

for $z \in \mathfrak{g}$. So

$$\sum_i T_{e_i}e_i = 0$$

and hence T is always of type $\mathcal{T}_2 \oplus \mathcal{T}_3$. Using the defining condition $\mathfrak{S}_{x,y,z} T_{xyz} = 0$ for the class $\mathcal{T}_2$ we obtain the equivalent condition

$$\mathfrak{S}_{x,y,z} g([x,y],z) = 0 \tag{8.3}$$

from (8.1). So $T \in \mathcal{T}_2$ if and only if $\lambda_1 + \lambda_2 + \lambda_3 = 0$. Making use of the classification given in [33] we conclude that in this case $\mathfrak{g}$ is the Lie algebra of $SL(2,\mathbb{R})$ or $E(1,1)$ (the group of rigid motions of the Minkowski 2-space). Further $T \in \mathcal{T}_3$ if and only if $T_{xxz} = 0$. This is equivalent to $\lambda_1 = \lambda_2 = \lambda_3$ and then $\mathfrak{g}$ is isomorphic to the Lie algebra of SU(2) (or SO(3)).

Next we consider the *non-unimodular case*. Here there exists an orthonormal basis (e_1,e_2,e_3) of $\mathfrak{g}$ such that

$$\begin{cases} [e_1,e_2] = \alpha e_2 + \beta e_3 , \\ [e_1,e_3] = \gamma e_2 + \delta e_3 , \\ [e_2,e_3] = 0 \end{cases} \tag{8.4}$$

where $\alpha + \delta = 2$. If the matrix

$$A = \begin{pmatrix} \alpha & \beta \\ \gamma & \delta \end{pmatrix}$$

is the unit matrix, (G,g) has constant negative curvature (see [33]). Since

$$\sum_i T_{e_i e_i} z = (\alpha + \delta) g(e_1, z) ,$$

T is never of type $\mathcal{T}_2 \oplus \mathcal{T}_3$. However if $\beta = \gamma$, T is of type $\mathcal{T}_1 \oplus \mathcal{T}_2$. Further $T \in \mathcal{T}_1 \oplus \mathcal{T}_3$ if and only if $T \in \mathcal{T}_1$ and this is the case if and only if $\alpha = \delta = 1$ and $\beta = \gamma = 0$. So A = I and then the result agrees with the property proved in chapter 5. If $\beta \neq \gamma$ we obtain a homogeneous structure T of general type $\mathcal{T}_1 \oplus \mathcal{T}_2 \oplus \mathcal{T}_3$ and T does not belong to a subclass.

Remark. The unimodular case E(1,1) can be obtained as a special case of the following group.

Let G be the three-dimensional matrix group given by

$$\begin{pmatrix} e^{-\alpha z} & 0 & 0 & e^{-\alpha z}x \\ 0 & e^{\beta z} & 0 & e^{\beta z}y \\ 0 & 0 & 1 & \frac{z}{2} \\ 0 & 0 & 0 & 1 \end{pmatrix} ,\alpha,\beta \in \mathbb{R}$$

with the left invariant metric

$$g = dz^2 + e^{2\alpha z}dx^2 + e^{-2\beta z}dy^2 .$$

The case $\alpha + \beta = 0$ corresponds to $\mathbb{H}^3$. In all the other cases the homogeneous structures T are of type $\mathcal{T}_1 \oplus \mathcal{T}_2$ and of type $\mathcal{T}_2$ if and only if $\alpha = \beta$. They never belong to $\mathcal{T}_1$ if $\alpha + \beta \neq 0$.

B. k-SYMMETRIC SPACES

For a second source of examples we consider the k-symmetric spaces and in particular the case k=3. We refer to [13],[29],[30] for the basic results and definitions.

It is well-known that a 3-symmetric space M has a canonical almost complex structure J which is compatible with the metric g. More precisely, (M,g,J) is an almost Hermitian manifold which is a *quasi-Kähler* manifold, i.e.

$$(\nabla_X J)Y + (\nabla_{JX} J)JY = 0 , \qquad X,Y \in \mathfrak{X}(M). \tag{8.5}$$

Further the (1,2)-tensor field T defined by

$$T_X Y = \frac{1}{2} J(\nabla_X J)Y , \qquad X,Y \in \mathfrak{X}(M) , \tag{8.6}$$

is a homogeneous structure on (M,g,J). (See for example [42].) We have

THEOREM 8.1. *Let* (M,g,J) *be a 3-symmetric manifold with canonical almost complex structure* J. *Then the homogeneous structure* T *given by* (8.6) *always belongs to* $\mathcal{T}_2 \oplus \mathcal{T}_3$. *Further,* $T \in \mathcal{T}_3$ *if and only if* (M,g,J) *is nearly Kählerian and* $T \in \mathcal{T}_2$ *if and only if* (M,g,J) *is almost Kählerian.*

Proof. From (8.5) and (8.6) it follows that

$$T_X Y + T_{JX} JY = 0 , \qquad X,Y \in \mathfrak{X}(M). \tag{8.7}$$

Further, let $(E_i, JE_i, i=1,\ldots,n)$ be an adapted orthonormal basis. Then (8.7) implies at once

$$\sum_{i=1}^{n} (T_{E_i} E_i + T_{JE_i} JE_i) = 0$$

and this means that $T \in \mathcal{T}_2 \oplus \mathcal{T}_3$.

Next $T \in \mathcal{T}_3$ if and only if $T_X X = 0$ for all $X \in \mathfrak{X}(M)$. (8.6) implies that this is equivalent to

$$(\nabla_X J)X = 0 , \quad X \in \mathfrak{X}(M) ,$$

and this is the defining relation for a *nearly Kähler* manifold.

Finally, the defining relation for the class $\mathcal{T}_2$ implies that $T \in \mathcal{T}_2$ if and only if

$$g((\nabla_X J)Y, JZ) + g((\nabla_Y J)JZ, X) + g((\nabla_{JZ} J)X, Y) = 0. \tag{8.8}$$

Here we used (8.6) and

$$(\nabla_X J)(JY) = - J(\nabla_X J)Y , \quad X,Y \in \mathfrak{X}(M) ,$$

together with the fact that M is quasi-Kählerian. Next we substitute Z by JZ in (8.8). We obtain at once

$$3(dF)(X,Y,Z) = g((\nabla_X J)Y,Z) + g((\nabla_Y J)Z,X) + g((\nabla_Z J)X,Y) = 0$$

where $F(X,Y) = g(JX,Y)$ for all $X,Y,Z \in \mathfrak{X}(M)$. This means that (M,g,J) is an *almost Kähler* manifold.

Note that the naturally reductive case, i.e. $T \in \mathcal{T}_3$, was already given in [13].

There exist 3-symmetric spaces which are not nearly Kählerian and not almost Kählerian. The following example is contained in [32]. Let G be the simply connected Lie group with left invariant metric and defined by the following Lie algebra g with inner product $\langle\, ,\, \rangle$: $(X_1,X_2,X_3,Y_1,Y_2,Y_3)$ is an orthonormal basis with respect to $\langle\, ,\, \rangle$ and

$$[X_1,X_2] = -[Y_1,Y_2] = -X_3 ,$$

$$[X_1,Y_2] = [Y_1,X_2] = Y_3 ,$$

$$[X_1,X_3] = [X_i,Y_3] = 0 , \quad i = 1,2,$$

$$[X_1,Y_1] = [X_2,Y_2] = [X_3,Y_3] = 0.$$

The group G is nilpotent and diffeomorphic to $\mathbb{R}^6$. It is in fact the group of type H considered in chapter 9. Further G is not naturally reductive and so not nearly Kählerian. Next put $U_j = X_j + iY_j$, $j = 1,2,3$ Then we have

$$3(dF)(U_1,U_2,U_3) = \underset{U_1,U_2,U_3}{\mathfrak{S}} g((\nabla_{U_1} J)U_2,U_3) \neq 0$$

and so G is not an almost Kähler manifold with respect to the canonical structure J which is associated with the symmetry $\{SU_j = e^{\frac{2\pi i}{3}} U_j,$ $j=1,2,3\}$.

There also exist almost-Kähler 3-symmetric spaces which are not nearly-Kählerian. First we note that any four-dimensional 3-symmetri space is almost Kählerian with respect to any invariant almost Hermitian structure. Now consider $\mathbb{R}^4(x,y,u,v)$ with the following metric :

$$g = \{-x + (x^2 + y^2 + 1)^{1/2}\}du^2 + \{x + (x^2 + y^2 + 1)^{1/2}\}dv^2$$
$$- 2ydudv + \lambda^2(1 + x^2 + y^2)^{-1}\{(1 + y^2)dx^2 + (1 + x^2)dy^2 - 2xydxdy\} ,$$

where $\lambda > 0$ (see [30, p. 137]). This is a 3-symmetric almost Kähler space which is not naturally reductive. Hence it is not nearly Kählerian (it is not Kählerian with respect to the canonical associated J but it is a Kähler manifold with respect to another almost complex structure [30, p. 87]).

A classical example of a nearly Kähler manifold is provided by S^6 with the almost complex structure induced by the Cayley numbers. So S^6 has at least two homogeneous structures : $T = 0$ corresponds to the representation $S^6 = SO(7)/SO(6)$ and the structure T given by (8.6) to $S^6 = G_2/SU(3)$ (see [43]).

Other examples may be found using the classification list in [13]. Of course it would also be interesting to consider the k-symmetric spaces in general but now we just consider the only k-symmetric space of dimension 3 (see [29]). This is the matrix group G, diffeomorphic to $\mathbb{R}^3$, given by

$$\begin{pmatrix} e^z & 0 & x \\ 0 & e^{-z} & y \\ 0 & 0 & 1 \end{pmatrix} \tag{8.9}$$

with the left invariant metric

$$g = e^{2z}dx^2 + e^{-2z}dy^2 + \lambda^2dz^2 , \quad \lambda > 0. \tag{8.10}$$

Here k is equal to 4. Using the same method as in chapter 7 and putting

$$\theta_1 = e^z dx , \quad \theta_2 = e^{-z}dy , \quad \theta_3 = \lambda dz , \tag{8.11}$$

it is easy to see that there exists *only one* homogeneous structure on this manifold (G,g). This is given by

$$T = -\frac{2}{\lambda}\theta_1 \otimes (\theta_1 \wedge \theta_3) + \frac{2}{\lambda}\theta_2 \otimes (\theta_2 \wedge \theta_3). \tag{8.12}$$

Hence T is of type $\mathcal{T}_2$. Further one finds that G is isomorphic to the semi-direct product of $\mathbb{R}$ and $\mathbb{R}^2$ (both with the additive group structure) and where the action of $\mathbb{R}$ on $\mathbb{R}^2$ is given by

$$\varphi(z) = (e^z x, e^{-z} y). \tag{8.13}$$

Following [33] we conclude that G is isomorphic to E(1,1) (see section A of this chapter). The connection $\tilde{\nabla} = \nabla - T$ is the canonical connection.

C. THE FOUR-DIMENSIONAL HYPERBOLIC SPACE

To decide about the inclusion relations we still need one example. Indeed, up to now we do not have a homogeneous structure T of type $\mathcal{T}_1 \oplus \mathcal{T}_3$ which is not of type $\mathcal{T}_1$ nor of type $\mathcal{T}_3$.

The results of chapter 6 imply that this example must have at least dimension 4. In what follows we show that the four-dimensional hyperbolic space provides such an example. Let $\mathbb{H}^4$ denote this space with the metric

$$ds^2 = r^2(y_1)^{-2} \sum_{i=1}^{4} (dy_i)^2 , \quad y_1 > 0 , \tag{8.14}$$

and put

$$\theta_1 = ry_1^{-1}dy_1 , \quad \theta_2 = ry_1^{-1}dy_2 , \quad \theta_3 = ry_1^{-1}dy_3 , \quad \theta_4 = ry_1^{-1}dy_4. \tag{8.15}$$

Next define T by

$$T_{XYZ} = g(X,Y)\varphi(Z) - g(X,Z)\varphi(Y) + \lambda(*\varphi)(X,Y,Z) \ , X,Y,Z \in \mathfrak{X}(\mathbb{H}^4) , \tag{8.16}$$

where λ is constant and

$$\varphi = \frac{1}{r}\theta_1 ; \tag{8.17}$$

$*$ denotes the Hodge operator.

To see that T is a homogeneous structure we put

$$(T_1)_{XYZ} = g(X,Y)\varphi(Z) - g(X,Z)\varphi(Y) , \quad X,Y,Z \in \mathfrak{X}(\mathbb{H}^4) , \qquad (8.18)$$

and define ξ by $g(\xi,Z) = \varphi(Z)$ for all $Z \in \mathfrak{X}(\mathbb{H}^4)$. Hence $\nabla_1 \xi = 0$ where $\nabla_1 = \nabla - T_1$ (see chapter 5). Then (8.16) and (8.17) imply that $T_X \xi = (T_1)_X \xi$ and hence

$$\tilde{\nabla}_X \xi = 0 , \quad X \in \mathfrak{X}(\mathbb{H}^4) , \qquad (8.19)$$

where $\tilde{\nabla} = \nabla - T$. It is clear that (8.19) is equivalent to $\tilde{\nabla}\varphi = 0$ and so we also have $\nabla(\ast\varphi) = 0$. This means that $\tilde{\nabla}T = 0$ and $T \in \mathcal{T}_1 \oplus \mathcal{T}_3$ but $T \notin \mathcal{T}_1$ and $T \notin \mathcal{T}_3$.

Without giving the detailed calculations we note that the Lie algebra $\mathfrak{g} = \mathfrak{m} \oplus \mathfrak{k}$ corresponding to this T is a semi-direct sum $\mathbb{R}^3 \oplus_\varphi \mathfrak{co}(3)$ where

$$\mathfrak{co}(3) = \begin{pmatrix} a & a_{12} & a_{13} \\ -a_{12} & a & a_{23} \\ -a_{13} & -a_{23} & a \end{pmatrix}, \quad a, a_{ij} \in \mathbb{R} .$$

$\mathbb{R}^3$ is the Abelian Lie algebra of dimension 3. Hence this homogeneous structure is associated with the representation

$$\mathbb{H}^4 = \mathbb{R}^3 \times_\varphi CO(3)/SO(3)$$

(see also [5, p. 45]).

From all these examples and from those given in chapters 4 to 7 we can conclude :

THEOREM 8.2. *The inclusion relations between the eight classes of homogeneous structures are all strict.*

D. REMARKS

a. It is possible to obtain several other examples of homogeneous structures by considering the direct product $M_1 \times M_2$ of two manifolds, where M_1 has a homogeneous structure T_1 and M_2 a homogeneous structure T_2.

b. An example which has been useful in another context (see [16]) is the following : Let S^3 be the three-dimensional sphere in $\mathbb{R}^4$ where $\mathbb{R}^4$ is regarded as the space of quaternions. Further let N denote the unit outward normal to S^3 and denote by φ_I, φ_J and φ_K the 1-forms on S^3 given by $\varphi_I(X) = \langle X, IN \rangle$, etc., where $X \in S^3$ and $\langle \, , \, \rangle$ denote the inner product of $\mathbb{R}^4$. Then the metric

$$g = \alpha^2\varphi_I^2 + \beta^2\varphi_J^2 + \gamma^2\varphi_K^2 ,$$

α,β,γ being constant, is a homogeneous metric on S^3. It is the standard metric on S^3 when $\alpha^2 = \beta^2 = \gamma^2$. The determination of the homogeneous structures and the corresponding groups is a nice exercise.

c. To have more *four-dimensional* examples we first recall a theorem of Jensen concerning four-dimensional Lie groups with a left-invariant Einstein metric [20].

<u>THEOREM</u>. *Let G be a four-dimensional Lie group with a left-invariant Riemannian metric. Then G is an Einstein space if and only if its Lie algebra g is one of the following solvable Lie algebras with the inner product defined, up to change in scale, by $X_1,\dots,X_4$ being an orthonormal basis. Distinct values of t define nonisomorphic Lie algebras.*

1. $[X_1,X_2] = 0$, $\quad [X_2,X_3] = 0$,

   $[X_1,X_3] = X_4$, $\quad [X_2,X_4] = 0$,

   $[X_1,X_4] = -X_3$, $\quad [X_3,X_4] = 0.$

As a Riemannian space this is flat.

2. $[X_1,X_2] = X_2 - tX_3$, $\quad [X_2,X_3] = 2X_4$,

   $[X_1,X_3] = tX_2 + X_3$, $\quad [X_2,X_4] = 0$,

$[X_1,X_4] = 2X_4$, $[X_3,X_4] = 0$, $0 \leqslant t < \infty$.

As a Riemannian space each of these is a Hermitian hyperbolic space with sectional curvature K *satisfying* $-1 \geqslant K \geqslant -4$.

3. $[X_1,X_2] = X_2$, $[X_2,X_3] = 0$,

$[X_1,X_3] = X_3 - tX_4$, $[X_2,X_4] = 0$,

$[X_1,X_4] = tX_3 + X_4$, $[X_3,X_4] = 0$, $0 \leqslant t < \infty$.

As a Riemannian space each of these is a real hyperbolic space with constant curvature K *equal to* -1.

4. $[X_1,X_2] = 0$, $[X_2,X_3] = 0$,

$[X_1,X_3] = X_3$, $[X_2,X_4] = X_4$,

$[X_1,X_4] = 0$, $[X_3,X_4] = 0$.

This Lie algebra is the direct sum of a two-dimensional Lie algebra with itself, and the Riemannian space is the direct product of a two-dimensional solvable group manifold, of constant curvature K *equal to* -1, *with itself.*

To determine a homogeneous structure T on these Lie groups we use the same method as for the three-dimensional Lie groups. More specifically we use (8.1). Then we obtain, denoting by θ_i the dual forms of X_i, $i=1,\dots,4$:

1. $$T = 2\theta_1 \otimes (\theta_3 \wedge \theta_4).$$

Hence $T \in \mathcal{T}_2 \oplus \mathcal{T}_3$.

2. $$T = 2\{\theta_2 \otimes (\theta_3 \wedge \theta_4 - \theta_1 \wedge \theta_2) - \theta_3 \otimes (\theta_1 \wedge \theta_3 + \theta_2 \wedge \theta_4) - \theta_4 \otimes (2\theta_1 \wedge \theta_4 + \theta_2 \wedge \theta_3)\} - 2t\,\theta_1 \otimes (\theta_2 \wedge \theta_3).$$

Here $T \in \mathcal{T}_1 \oplus \mathcal{T}_2 \oplus \mathcal{T}_3$.

3. $$T = -2\{\theta_2 \otimes (\theta_1 \wedge \theta_2) + \theta_3 \otimes (\theta_1 \wedge \theta_3) + \theta_4 \otimes (\theta_1 \wedge \theta_4)\}$$
$$- 2t\theta_1 \otimes (\theta_3 \wedge \theta_4).$$

In this case $T \in \mathcal{T}_1 \oplus \mathcal{T}_2 \oplus \mathcal{T}_3$ and $T \in \mathcal{T}_1$ if and only if $t = 0$.

4. $$T = -2\{\theta_3 \otimes (\theta_1 \wedge \theta_3) + \theta_4 \otimes (\theta_2 \wedge \theta_4)\}.$$

Here $T \in \mathcal{T}_1 \oplus \mathcal{T}_2$.

Note that in the cases 1 and 3, X_2, X_3, X_4 span a three-dimensional Abelian subalgebra. Hence $\mathfrak{g}$ is a semi-direct sum of $\mathbb{R}$ and $\mathbb{R}^3$. In the case 2, X_2, X_3 and X_4 span a Lie algebra isomorphic to the Heisenberg algebra $\mathfrak{h}$ of dimension 3. Hence $\mathfrak{g}$ is a semi-direct sum of $\mathbb{R}$ and $\mathfrak{h}$. Finally, in case 4, we put

$$\begin{cases} X_1 = e_4 - \frac{1}{2} e_3, & X_2 = e_4 + \frac{1}{2} e_3, \\ X_3 = e_1 + e_2, & X_4 = e_1 - e_2. \end{cases}$$

Then e_1, e_2 and e_3 span a subalgebra isomorphic to the Lie algebra of $E(1,1)$. Therefore $\mathfrak{g}$ is a semi-direct product of $\mathbb{R}$ and this Lie algebra.

9. GENERALIZED HEISENBERG GROUPS

In chapter 7 we considered in detail the Heisenberg group. It is well-known that this group plays an important role in physics, for example in quantum mechanics and in the theory of contact transformations. It is also used extensively in harmonic analysis. Sometimes it provides nice examples in Riemannian geometry. We refer to [16],[31] where the Heisenberg group is used in connection with the problem of characterizing spaces by means of the volume of small geodesic spheres. See also [40].

Two properties of the Heisenberg group are important to be noted here. On the one hand it is an example of a 2-step nilpotent group. On the other hand we have shown in Theorem 7.2 that it is a naturally reductive homogeneous space. In this chapter we first show how the first property leads to a nice generalization, namely to the so-called *generalized Heisenberg groups* or *groups of type* H (see [21],[22],[23],[50]). But we will also show that some of the properties of the Heisenberg group do not hold for this larger class. More precisely we will show that there are groups of type H which are not naturally reductive. To do this we mainly concentrate on a remarkable six-dimensional example of Kaplan [23] but we also provide a different proof using only the methods of these notes. Finally we show that this six-dimensional manifold provides an example for some open problems related to the work of D'Atri and Nickerson [9],[10],[11].

A. LIE GROUPS OF TYPE H

First we give a brief survey on some general aspects of groups of type H. We refer to [21],[22],[23] for more details. At the same time we concentrate on the naturally reductive case and we give a different proof of the main theorem using the theory of two-fold vector cross products.

First we start with the definition of such a group. Let V and Z be two real vector spaces of dimension n and m, $m \geq 1$, both equipped with an inner product which we shall denote for both spaces by the same symbol $< , >$. Further let $j : Z \to \mathrm{End}(V)$ be a linear map such that

$$|j(a)x| = |x|\,|a| , \qquad x \in V , a \in Z , \tag{9.1}$$

$$j(a)^2 = -|a|^2 I , \qquad a \in Z. \tag{9.2}$$

Using polarization we obtain from (9.1)

$$< j(a)x , j(b)x > = < a,b > |x|^2 , \tag{9.3}$$

$$< j(a)x , j(a)y > = |a|^2 < x,y > , \tag{9.4}$$

for all $x,y \in V$ and $a,b \in Z$.

Next we define the Lie algebra $\mathfrak{n}$ as the direct sum of V and Z together with the bracket defined by

$$[a + x,b + y] = [x,y] \in Z , \tag{9.5}$$

$$< [x,y],a > = < j(a)x,y > \tag{9.6}$$

where $a,b \in Z$ and $x,y \in V$. Then $\mathfrak{n}$ is said to be a *Lie algebra of type* H. It is a 2-step nilpotent Lie algebra with center Z.

The simply connected, connected Lie group N whose Lie algebra is $\mathfrak{n}$ is called a *Lie group of type* H or a *generalized Heisenberg group*. There are infinitely many groups of type H with center of any given dimension. Further we note that the Lie algebra $\mathfrak{n}$ has an inner product such that V and Z are orthogonal :

$$< a + x,b + y > = < a,b > + < x,y > ,$$

$a,b \in Z$ and $x,y \in V$. Hence the Lie group has a left invariant metric induced by this metric on $\mathfrak{n}$.

As we will see, some special Lie algebras of type H will play a fundamental role in what follows. These non-Abelian algebras can be obtained using composition algebras W (the complex numbers $\mathbb{C}$, the quaternions $\mathbb{H}$ and the Cayley numbers Cay) as follows : Let Z be the subspace of W formed by the purely imaginary elements. Further let $V = W^n$, $n \in \mathbb{N}_0$, and put $j : Z \to \mathrm{End}(V)$ for the linear map defined by

$$j(a)x = ax \tag{9.7}$$

where ax denotes the ordinary scalar multiplication of a and x. The corresponding groups are the *Heisenberg groups* or their *quaternionic* and *Cayley analogs*.

Next we look for the naturally reductive groups. The result has been proved in a different way by Kaplan in [23]. Here we will use Theorem 6.1.

THEOREM 9.1. *The homogeneous manifold* $(N, \langle , \rangle)$ *is naturally reductive if and only if* N *is a Heisenberg group or a quaternionic analog.*

To prove this we first prove

LEMMA 9.2. *If* $(N, \langle , \rangle)$ *is naturally reductive, then* $\dim Z = 1$ *or* 3.

Proof. It follows from Theorem 6.1 that there exists a tensor T of type (1,2) such that

$$T_X Y + T_Y X = 0$$

and which satisfies the conditions (AS). Further let ρ denote the Ricci tensor of the manifold $(N, \langle , \rangle)$. Then we obtain from ((AS)(ii)) by contraction

$$(\nabla_X \rho)_{YZ} = - \rho_{T_X YZ} - \rho_{Y T_X Z}. \tag{9.8}$$

The connection of Levi Civita has been computed in [22]. We have

$$\left\{ \begin{array}{l} \nabla_x y = \frac{1}{2}[x,y] , \\ \nabla_a x = \nabla_x a = - \frac{1}{2} j(a)x , \end{array} \right. \tag{9.9}$$

$$\left| \nabla_a b = 0 \right. \tag{9.9}$$

where $x,y \in V$ and $a,b \in Z$. For the Ricci tensor one obtains (see [22]) :

$$\left\{ \begin{array}{l} \rho_{xy} = -\frac{m}{2} < x,y > , \\ \rho_{ab} = \frac{n}{4} < a,b > , \\ \rho_{xa} = 0. \end{array} \right. \tag{9.10}$$

Then it follows easily from (9.9) and (9.10) that all the complements of $\nabla\rho$ vanish except

$$(\nabla_x \rho)_{bz} = -\frac{n+2m}{8} < j(b)x,z > . \tag{9.11}$$

Hence (9.8) will be satisfied if and only if

$$< T_x z \ , \ b > = \frac{1}{2} < j(b)x,z > , \tag{9.12}$$

$$< T_a z \ , \ b > = 0. \tag{9.13}$$

Since T_u is skew-symmetric for all $u \in n$ we must have

$$T_a b \in Z , \tag{9.14}$$

$$T_x b = - T_b x = - \frac{1}{2} j(b)x. \tag{9.15}$$

Next we put $\tilde{\nabla} = \nabla - T$. Then it follows from (9.9) and (9.15) that

$$\left\{ \begin{array}{l} \tilde{\nabla}_a x = - j(a)x , \\ \tilde{\nabla}_a b = - T_a b. \end{array} \right. \tag{9.16}$$

So we obtain

$$< (\tilde{\nabla}_a T)_{yz}, b > = \frac{1}{2} \{ < j(b)j(a)y,z > + < j(b)y,j(a)z > \tag{9.17}$$

$$+ < j(T_a b)y,z >\} .$$

Since j(a) is skew-symmetric for all $a \in Z$, $\tilde{\nabla}T = 0$ implies from (9.17) :

$$j(T_a b) = j(a)j(b) - j(b)j(a). \tag{9.18}$$

Next, further polarization of (9.2) gives

$$j(a)j(b) + j(b)j(a) = - 2 < a,b > I \tag{9.19}$$

and hence (9.18) becomes

$$j(T_a b) = 2\{j(a)j(b) + < a,b > I\} . \tag{9.20}$$

Finally we have

$$j(T_a b)^2 = - |T_a b|^2 \, I$$

and from this, using (9.19) and (9.20), we conclude that

$$|T_a b|^2 = 4(|a|^2|b|^2 - < a,b >^2). \tag{9.21}$$

But we also have

$$< T_a b,a > = < T_a b,b > = 0. \tag{9.22}$$

So, putting

$$\tau_a b = \frac{1}{2} T_a b \quad , \quad a,b \in Z,$$

we can conclude from (9.21) and (9.22) that τ is a two-fold vector cross product on Z (see [7]). Hence we must have dim Z = 1 if $T_a b = 0$ or otherwise dim Z = 3 or 7.

It remains to prove that dim Z = 7 is not possible. To prove this we first recall that with $W = \mathbb{R} \oplus Z$ and the multiplication

$$1a = a1 = a \,, \qquad\qquad 1 \in \mathbb{R},$$

$$ab = \tau_a b - \langle a,b \rangle 1 , \quad a,b \in Z ,$$

we obtain an 8-dimensional composition algebra. The inner product on Z can be extended to W by putting

$$|1| = 1$$

and taking Z to be orthogonal to $\mathbb{R}$. Then (9.21) implies that W is associative. Indeed, let $\tilde{j} : W \to \mathrm{End}(V)$ be the linear map defined by

$$\tilde{j}(1) = I , \quad \tilde{j}(a) = j(a) \text{ for } a \in Z . \tag{9.23}$$

It is clear that $\tilde{j}$ is injective. Moreover $\tilde{j}(ab) = \tilde{j}(a)\tilde{j}(b)$ since

$$\begin{aligned}\tilde{j}(ab) = \tilde{j}(\tau_a b) - \langle a,b \rangle I &= \frac{1}{2}\tilde{j}(T_a b) - \langle a,b \rangle I \\ &= j(a)j(b)\end{aligned}$$

as follows from (9.20) and (9.23). Hence $\tilde{j}$ is a monomorphism between the algebras W and End(V) and so W is associative. But since any 8-dimensional composition algebra is not associative, this excludes the case dim Z = 7. Hence the lemma is proved.

<u>Proof of Theorem 9.1.</u> Using the classification of Clifford modules, Kaplan proved in [23] that for dim Z = 1 the corresponding groups N are the Heisenberg groups and for dim Z = 3 the groups N are the quaternionic analogs.

So to finish the proof we have to show that in these cases there exists a tensor T satisfying the required conditions. Therefore, let T be defined as follows :

$$\begin{cases} T_x y = - T_y x = \frac{1}{2}[x,y] , \\ T_x a = - T_a x = - \frac{1}{2} j(a)x = - \frac{1}{2} ax , \\ T_a b = 2\{ab + \langle a,b \rangle 1\}. \end{cases} \tag{9.24}$$

Using the explicit expression for R given in [22] and the properties of the composition algebras (see [7]) it is not difficult to verify that the T defined by (9.24) satisfies the conditions (AS), or equivalently, the conditions $\tilde{\nabla}R = \tilde{\nabla}T = 0$.

B. GEODESICS AND KILLING VECTOR FIELDS ON GROUPS OF TYPE H

Now we discuss the following question : When are the geodesics on a group $(N, \langle , \rangle)$ of type H orbits of one-parameter subgroups of isometries of $(N, \langle , \rangle)$? To do this we do not use the description of the full group of isometries of $(N, \langle , \rangle)$ given in [22] (see also [54]), but instead consider the Killing vector fields.

First we determine a global coordinate system $(v_1,\ldots,v_n;u_1,\ldots,u_m)$ on N. To do this, let $(x_1,\ldots,x_n)$ and $(a_1,\ldots,a_m)$ be orthonormal frames on V and Z. Then we put for $p \in N$:

$$\begin{cases} v_i(p) = v_i(\exp(x(p) + a(p)) = \langle x(p), x_i \rangle \ , \ i = 1,\ldots,n, \\ u_\alpha(p) = u_\alpha(\exp(x(p) + a(p)) = \langle a(p), a_\alpha \rangle \ , \ \alpha = 1,\ldots,m. \end{cases} \tag{9.25}$$

Then we have

$$\begin{cases} \dfrac{\partial}{\partial v_i} = x_i - \dfrac{1}{2} \sum\limits_{\alpha,j} A^{\alpha}_{ji} v_j a_\alpha \ , \\ \dfrac{\partial}{\partial u_\alpha} = a_\alpha \ , \end{cases} \tag{9.26}$$

where the A^{α}_{ji} are the structure constants of the Lie algebra $\mathfrak{n}$, i.e.

$$[x_i, x_j] = \sum_\alpha A^{\alpha}_{ij} a_\alpha \ . \tag{9.27}$$

Next, let A, respectively B, be a skew-symmetric endomorphism of V, respectively Z, such that

$$Aj(a) - j(a)A = j(B(a)) \ , \qquad a \in Z \ , \tag{9.28}$$

and put

$$A(x_i) = \sum_j a_{ji} x_j \quad , \quad B(a_\alpha) = \sum_\beta b_{\beta\alpha} a_\beta \ .$$

Then we have

<u>THEOREM 9.3.</u> *The Killing vector fields* ξ *of* $(N, <, >)$ *are given by*

$$\xi = \sum_i \xi_i \frac{\partial}{\partial v_i} + \sum_\alpha \xi_\alpha \frac{\partial}{\partial u_\alpha}$$

where

$$\xi_i = \sum_j a_{ij} v_j + \lambda_i \,, \qquad \lambda_i = \text{const.} \,, \tag{9.29}$$

$$\xi_\alpha = \sum_\beta b_{\alpha\beta} u_\beta + \frac{1}{2} \sum_{i,j} A^\alpha_{ij} v_i \lambda_j + \mu_\alpha \,, \qquad \mu_\alpha = \text{const.} \tag{9.30}$$

<u>Proof</u>. The Killing equations can be written as follows :

$$\begin{cases} g(\nabla_{x_i} \xi, x_j) + g(\nabla_{x_j} \xi, x_i) = 0 \,, \\ g(\nabla_{a_\alpha} \xi, a_\beta) + g(\nabla_{a_\beta} \xi, a_\alpha) = 0 \,, \\ g(\nabla_{a_\alpha} \xi, x_i) + g(\nabla_{x_i} \xi, a_\alpha) = 0 \,. \end{cases} \tag{9.31}$$

Next let ρ be the Ricci tensor of $(N, <, >)$. Then the Lie derivative $\mathcal{L}_\xi$ vanishes. More specifically we have

$$\rho([\xi, a_\alpha], x_i) + \rho([\xi, x_i], a_\alpha) = 0. \tag{9.32}$$

Using (9.9) and (9.10) we derive the following conditions which are equivalent to (9.31) and (9.32) :

$$\begin{cases} x_i(\xi_j) + x_j(\xi_i) = 0 \,, \\ a_\alpha(\tilde{\xi}_\beta) + a_\beta(\tilde{\xi}_\alpha) = 0 \,, \\ a_\alpha(\xi_i) = 0 \,, \\ x_i(\tilde{\xi}_\alpha) + \sum_h \xi_h < [x_i, x_h] \,, a_\alpha > = 0 \,, \end{cases} \tag{9.33}$$

where

$$\tilde{\xi}_\alpha = \xi_\alpha - \frac{1}{2} \sum_{i,j} A^\alpha_{ji} v_j \xi_i \ . \tag{9.34}$$

From the first and third condition in (9.33) we derive

$$\frac{\partial \xi_i}{\partial u_\alpha} = 0 \ , \qquad \frac{\partial \xi_i}{\partial v_j} + \frac{\partial \xi_j}{\partial v_i} = 0$$

and hence

$$\xi_i = \sum_j a_{ij} v_j + \lambda_i \ , \tag{9.35}$$

where $a_{ij} + a_{ji} = 0$ and a_{ij}, λ_i are constants. Similarly, the second condition in (9.33) gives

$$\tilde{\xi}_\alpha = \sum_\beta b_{\alpha\beta}(v) u_\beta + \eta_\alpha(v) \tag{9.36}$$

with $b_{\alpha\beta} + b_{\beta\alpha} = 0$. Next we determine the functions $b_{\alpha\beta}$ and η_α using the last equation of (9.33). Therefore we substitute (9.36) in the equation. Differentiation with respect to u_β gives that the $b_{\alpha\beta}$ are constant. Moreover

$$\frac{\partial \eta_\alpha}{\partial v_i} + \frac{1}{2} \sum_\beta \left(\sum_j A^\beta_{ji} v_j\right) b_{\alpha\beta} + \sum_{h,j} A^\alpha_{ih} a_{hj} v_j = 0 \ . \tag{9.37}$$

The integrability conditions of this system are

$$\sum_\beta A^\beta_{ij} b_{\alpha\beta} - \sum_h A^\alpha_{ih} a_{hj} + \sum_h A^\alpha_{jh} a_{hi} = 0 \ . \tag{9.38}$$

Taking into account (9.27) and (9.6), we find that (9.38) is equivalent to (9.28).

Conversely, suppose we have (9.28). Then we have

$$\eta_\alpha = -\frac{1}{2} \sum_{i,j,h} A^\alpha_{ih} a_{hj} v_i v_j + \mu_\alpha \ , \qquad \mu_\alpha = \text{const.} \tag{9.39}$$

So the required formula (9.30) follows from (9.39), (9.34) and (9.36).

The geodesics of the manifold $(N, \langle , \rangle)$ have been calculated explicitly in [22], [23]. Let $\gamma(t) = \exp(x(t) + a(t))$ be the geodesic tangent at 0 to the vector $\dot{\gamma}(0) = \lambda + \mu$, $\lambda \in V$, $\mu \in Z$. Then we have

$$\begin{cases} x(t) = \dfrac{1 - \cos|\mu|t}{|\mu|^2} j(\mu)\lambda + \dfrac{\sin|\mu|t}{|\mu|} \lambda , \\ a(t) = \left\{ t + \dfrac{|\lambda|^2}{2|\mu|^2} \left(t - \dfrac{\sin|\mu|t}{|\mu|}\right) \right\} \mu \end{cases} \tag{9.40}$$

for $\mu \neq 0$ and

$$\begin{cases} x(t) = t\lambda , \\ a(t) = 0 \end{cases} \tag{9.41}$$

for $\mu = 0$.

From this it is clear that if $\mu = 0$, $\gamma(t) = (\exp t\lambda)0$ is the orbit of a one-parameter subgroup of isometries of $(N, \langle , \rangle)$. More generally we have

THEOREM 9.4. *The geodesic* $\gamma(t)$, *with* $\dot{\gamma}(0) = \lambda + \mu$, *is an orbit of a one-parameter subgroup of isometries of* $(N, \langle , \rangle)$ *if and only if there exist skew-symmetric endomorphisms* A *and* B *of* V *and* Z *such that*

$$\begin{cases} A(\lambda) = j(\mu)\lambda , \\ B(\mu) = 0 , \\ Aj(a) - j(a)A = j(B(a)) \end{cases} \tag{9.42}$$

for all $a \in Z$.

Proof. The conditions (9.42) are the necessary and sufficient conditions for the existence of a Killing vector field ξ such that $\xi(\gamma(t)) = \dot{\gamma}(t)$ for all t.

C. THE GEOMETRY OF THE SIX-DIMENSIONAL GROUP OF TYPE H

Now we use the results of 9.A and 9.B to study a particular example which has been discussed by Kaplan in [23]. It will follow from the explicit expression of the Lie algebra (9.57) that there is only one group of type H of dimension six. This example can be described as follows. Let $V = \mathbb{H}$, the space of quaternions, and let Z be a two-dimensional subspace of purely imaginary quaternions. Further, let $j : Z \to \mathrm{End}(V)$ be the linear map defined by

$$j(a)x = ax \ , \qquad a \in Z \ , \ x \in V \ , \tag{9.43}$$

i.e. $j(a)x$ is the ordinary multiplication of x by a. It is not difficult to prove that $n = V \oplus Z$ is a Lie algebra of type H. Further it follows from Theorem 9.1 that the corresponding Lie group N of type H is a homogeneous space which is not naturally reductive.

In [23] Kaplan proved that the geodesics of this group N are all orbits of one-parameter subgroups of isometries. Now we shall give a new proof of this result using Theorem 9.4.

THEOREM 9.5. *Let* $(N, \langle\, , \rangle)$ *denote the six-dimensional group of type* H. *Then the geodesics of* $(N, \langle\, , \rangle)$ *are orbits of one-parameter subgroups of isometries of* $(N, \langle\, , \rangle)$.

Proof. Let $\gamma(t)$ be the geodesic of N through $\gamma(0) = 0$ and such that $\gamma(0) = \lambda + \mu$, where $\lambda \in V$, $\mu \in Z$. First we suppose $\dot{\lambda} \neq 0$ and $\mu \neq 0$. Let (a_1, a_2) be an orthonormal frame of Z such that $\mu = \mu_1 a_1$ and let x_1 be a unit vector of V such that $\lambda = \lambda_1 x_1$. Then $\{x_1, j(a_1)x_1, j(a_2)x_1, j(a_1)j(a_2)x_1\}$ is an orthonormal basis of V.

It follows from Theorem 9.4 that there exists a Killing vector field ξ such that $\xi(\gamma(t)) = \dot{\gamma}(t)$ if and only if there exist a skew-symmetric endomorphism A of V and a skew-symmetric endomorphism B of Z satisfying (9.42). Now it is clear that in our case $B = 0$ and further A is uniquely determined by

$$\begin{cases} A(x_1) = j(a_1)x_1 \ , & A(j(a_1)x_1) = -\,x_1 \ , \\ A(j(a_2)x_1) = -\,j(a_1)j(a_2)x_1 \ , & A(j(a_1)j(a_2)x_1) = j(a_2)x_1 . \end{cases} \tag{9.44}$$

Hence these geodesics are orbits of *unique* one-parameter subgroups of isometries.

If $\lambda = 0$ or $\mu = 0$ we have $\gamma(t) = \exp t\mu$ or $\gamma(t) = \exp t\lambda$. Hence we put $A = B = 0$. In this case the geodesics are again orbits of one-parameter subgroups but these subgroups are *not uniquely* determined.

REMARK. It is important to note that this property cannot be extended to all groups of type H. Indeed, Kaplan proved in [23] that the groups of type H with $m \equiv 0 \pmod 4$ are such that not all the geodesics are orbits of one-parameter subgroups.

In what follows we want to concentrate on another property which holds at least partly for all groups of type H. It was this property which was the starting point for our research concerning the Ambrose-Singer theorem. Before doing this we first need some preliminaries.

Let (M,g) be an n-dimensional Riemannian manifold and m a point of M. Further, let $(x_1,\ldots,x_n)$ be a system of normal coordinates centered at m and p a point of M such that $r = d(m,p) < i(m)$ where $i(m)$ is the injectivity radius at m. Then p can be joined to m by a unique geodesic γ. Put $\gamma(0) = m$ and $\gamma(r) = p = \exp_m(r\xi)$ where ξ is the unit velocity vector. The *geodesic symmetry* ψ (about m) is defined by

$$\psi_m : M \to M , \quad p = \exp_m(r\xi) \mapsto \psi_m(p) = \exp_m(-r\xi) = -p$$

and this is an involutive local diffeomorphism.

Riemannian manifolds with *volume-preserving* or, equivalently, *divergence-preserving geodesic symmetries* were studied in [9],[10],[11] and such manifolds are called *D'Atri spaces* in [51],[52]. They can be characterized as follows. Let

$$\theta = (\det g_{ij})^{1/2} = (\det g(\frac{\partial}{\partial x_i} , \frac{\partial}{\partial x_j}))^{1/2}.$$

Then it is not difficult to see that (M,g) is a D'Atri space if and only if

$$\theta(-p) = \theta(p)$$

for all $m \in M$ and all p near m. So θ has *antipodal symmetry*. For other characterizations we refer to [9],[51],[52].

Examples of D'Atri spaces are the so-called *commutative spaces*. We refer to [51] for the definition and further results. This last class of manifolds includes the *harmonic spaces* and products of such spaces. Also all symmetric spaces are commutative. All naturally reductive homogeneous Riemannian manifolds are D'Atri spaces and these were, among the homogeneous manifolds, the only previously known examples.

This property of naturally reductive homogeneous spaces was proved in [11]. A different proof is given in [9]. We state this last result of [11] since it shows how it was just this theorem which led us to take a closer look at the theorem of Ambrose and Singer.

<u>THEOREM 9.6.</u> *Let* M *be a real analytic Riemannian manifold which has, in a neighbourhood of each point, a* C^∞ *tensor field* T *of type* (1,2) *satisfying*

$$\begin{cases} (\nabla_X R)_{XY}X = T_X R_{XY}X - R_{XT_XY}X \,, \\ (\nabla_X T)_X = 0 \,. \end{cases} \tag{9.45}$$

Then the geodesic symmetries are locally divergence-preserving.

Using the equations (AS) and the results of chapter 6 we easily see that indeed every naturally reductive space is a D'Atri space. It was the search for non-naturally reductive examples which gave rise to our work (see also [51],[52]) and to part of the work of Kaplan [23]. Note that up to now no examples of nonhomogeneous D'Atri spaces are known.

In [23] Kaplan proved the following remarkable result :

<u>THEOREM 9.7.</u> *All the groups of type* H *are D'Atri spaces.*

It follows from this that the volume-preserving geodesic symmetry property does not characterize the naturally reductive spaces among the homogeneous spaces. Moreover we do not know if a D'Atri space is necessarily locally homogeneous.

But D'Atri proved in [9] a stronger result. Let $G = (g_{ij})$ be the matrix of g with respect to a normal coordinate system at m. Then the eigenvalues of G are independent of the choice of the normal coordinate system at m. D'Atri proved that *all* these eigenvalues have the antipodal symmetry when the manifold is naturally reductive. A different proof for this result can be given using the special form of the Jacobi equation in terms of the canonical connection and the associated tensor T or the torsion tensor of this connection. Now we shall prove that the six-dimensional group of type H also has this property so that again this cannot be characteristic for the class of naturally reductive homogeneous spaces.

THEOREM 9.8. *Let* $(N,<,>)$ *be the six-dimensional group of type* H. *Then all the eigenvalues of the matrix of the metric* $<,>$ *with respect to any normal coordinate system have the antipodal symmetry.*

Proof. Let G be the matrix of $<,>$ with respect to a normal coordinate system centered at 0 and let γ be a geodesic through 0 and $q \in \gamma$. Then we have to prove that

$$\det(G - \lambda I)(-q) = \det(G - \lambda I)(q).$$

Therefore we have to compute G(q). We put t = 1 in (9.40) and (9.41). This gives the relation between the global coordinates (v_i, u_α) of $p = \exp(x(1) + a(1))$ and the normal coordinates (λ_i, μ_α) with respect to the basis $(x_i(0), a_\alpha(0))$ of T_0N. Further, from (9.26) we obtain that in general the dual frame of (x_i, a_α) is determined by the left invariant 1-forms θ_i, ψ_α where

$$\theta_i = dv_i \, , \tag{9.46}$$

$$\psi_\alpha = du_\alpha - \frac{1}{2} \sum_{j,i} A^\alpha_{ji} v_j dv_i \tag{9.47}$$

for i,j = 1,...,n and α = 1,...,m. Hence

$$<,> = \sum_i \theta_i^2 + \sum_\alpha \psi_\alpha^2 \, . \tag{9.48}$$

To obtain the components of G(p) with respect to normal coordinates we have to express dv_i and du_α as functions of $d\lambda_i$, $d\mu_\alpha$. Therefore we use the fact that the eigenvalues of G are independent of the normal coordinate system choosen at 0 and hence we put

$$\lambda = \lambda_1 x_1 \quad , \quad \mu = \mu_1 a_1 .$$

On the six-dimensional manifold we then choose a basis (a_1,a_2) for Z and the basis (x_1,x_2,x_3,x_4) of V with

$$\begin{cases} x_2 = j(a_1)x_1 \ , \\ x_3 = j(a_2)x_1 \ , \\ x_4 = j(a_1)j(a_2)x_1 . \end{cases} \tag{9.49}$$

An easy calculation now shows that

$$\begin{cases} \psi_1 = du_1 - \frac{1}{2}(v_1 dv_2 - v_2 dv_1) - \frac{1}{2}(v_3 dv_4 - v_4 dv_3) \ , \\ \psi_2 = du_2 - \frac{1}{2}(v_1 dv_3 - v_3 dv_1) + \frac{1}{2}(v_2 dv_4 - v_4 dv_2) . \end{cases} \tag{9.50}$$

Moreover, from (9.40) and (9.41) we derive

$$\begin{cases} v_1 = -\beta(\mu_1\lambda_2 + \mu_2\lambda_3) + \alpha\lambda_1 \ , \\ v_2 = \beta(\mu_1\lambda_1 + \mu_2\lambda_4) + \alpha\lambda_2 \ , \\ v_3 = -\beta(\mu_1\lambda_4 - \mu_2\lambda_1) + \alpha\lambda_3 \ , \\ v_4 = \beta(\mu_1\lambda_3 - \mu_2\lambda_2) + \alpha\lambda_4 \ , \\ u_1 = \gamma\mu_1 \ , \\ u_2 = \gamma\mu_2 \ , \end{cases} \tag{9.51}$$

where

$$\begin{cases} \alpha = \dfrac{\sin|\mu|}{|\mu|} \\[2ex] \beta = \dfrac{1 - \cos|\mu|}{|\mu|} \, , \\[2ex] \gamma = 1 + \dfrac{|\lambda|^2}{2|\mu|^2}\left(1 - \dfrac{\sin|\mu|}{|\mu|}\right) \end{cases} \tag{9.52}$$

if $|\mu| \neq 0$. The case $|\mu| = 0$ can be obtained by continuity.

Note that

$$\alpha(-p) = \alpha(p) \ , \quad \beta(-p) = \beta(p) \ , \quad \gamma(-p) = \gamma(p)$$

and

$$\begin{cases} d\alpha_{|p} = A(p)d\mu_{1|p} \, , \\[1ex] d\beta_{|p} = B(p)d\mu_{1|p} \, , \\[1ex] d\gamma_{|p} = C(p)d\mu_{1|p} + D(p)d\lambda_{1|p} \end{cases} \tag{9.53}$$

where

$$A(-p) = -A(p) \ , \qquad B(-p) = -B(p) \ ,$$

$$C(-p) = -C(p) \ , \qquad D(-p) = -D(p).$$

Now from (9.51) we compute dv_i, du_α; then use (9.46) and (9.47) and substitute in (9.48) after evaluating at p. So we obtain the following expression for the characteristic polynomial :

$$P_p(\lambda) = \det(G-\lambda I)(p) = \begin{vmatrix} e_1-\lambda & 0_1 & 0 & 0 & e_7 & 0 \\ 0_1 & e_2-\lambda & 0 & 0 & 0_3 & 0 \\ 0 & 0 & e_3-\lambda & 0_2 & 0 & 0_4 \\ 0 & 0 & 0_2 & e_4-\lambda & 0 & e_8 \\ e_7 & 0_3 & 0 & 0 & e_5-\lambda & 0 \\ 0 & 0 & 0_4 & e_8 & 0 & e_6-\lambda \end{vmatrix} \tag{9.54}$$

where

$$e_j(-p) = e_j(p) \quad , \quad O_\beta(-p) = - O_\beta(p) \tag{9.55}$$

for $j = 1,\ldots,8$ and $\beta = 1,\ldots,4$.

From this it is clear that

$$P_{-p}(\lambda) = P_p(\lambda)$$

if $O_\beta = 0$ for all β. If at least one O_β is different from zero, the same result is obtained as can be seen by multiplying the first, third and fifth row by O_β.

This finishes the proof since p can be any arbitrary point q near 0.

REMARKS

a. As we mentioned already we do not know of any nonhomogeneous manifold having the properties mentioned in Theorems 9.7 and Theorem 9.8. It would be nice to know if a Riemannian manifold with one of these properties is (locally) homogeneous since this would imply that a harmonic space is (locally) homogeneous.

Further it would also be of some interest to know if the property of Theorem 9.8 has something to do with the fact that all geodesics are orbits of one-parameter subgroups of isometries. Is it possible to extend Theorem 9.8 to the class of manifolds whose geodesics are all orbits of one-parameter subgroups of isometries ?

b. In [1] Ambrose and Singer also considered another class of homogeneous spaces, namely the class such that

$$\nabla_X T_X = 0 \tag{9.56}$$

for all $X \in \mathfrak{X}(M)$. Note that the condition $T_X X = 0$ implies (9.56) as can be seen from ((AS)(iii)).

It is straightforward but tedious to prove that a homogeneous structure T satisfying (9.56) cannot exist on the six-dimensional group of type H.

D. SOME FURTHER RESULTS

Finally we consider some other properties of the six-dimensional group of type H in relation with the theory of k-symmetric spaces.

First we write down explicitly the brackets for the Lie algebra $\mathfrak{n}$ of this group $(N, \langle\,,\rangle)$. We obtain easily with the notations of above

$$\begin{cases} [x_1,x_2] = a_1\,, & [x_1,x_3] = a_2\,, \\ [x_2,x_4] = -\,a_2\,, & [x_3,x_4] = a_1\,, \\ \text{all the other brackets being zero.} & \end{cases} \tag{9.57}$$

Next we put

$$\begin{cases} U_1 = x_1 + ix_4\,, \\ U_2 = x_2 + ix_3\,, \\ U_3 = -a_1 + ia_2\,, \end{cases} \tag{9.58}$$

and define the linear map S of $\mathfrak{n}$ by

$$SU_j = e^{\frac{2\pi i}{3}}\,U_j\,, \qquad j = 1,2,3.$$

It follows at once from (9.57) and (9.58) that S is an isometric automorphism of the Lie algebra $(\mathfrak{n}, \langle\,,\rangle)$ and moreover $S^3 = I$. Hence N is a *3-symmetric space*. Note that the canonical almost complex structure J associated with S, i.e.

$$S = -\frac{1}{2}\,I + \frac{\sqrt{3}}{2}\,J\,,$$

is neither nearly Kähler nor almost Kähler. The fact that J is not nearly Kähler agrees with the property that $(N, \langle\,,\rangle)$ is not naturally reductive.

Next consider the linear map S defined by

$$SU_1 = iU_1 , \quad SU_2 = iU_2 , \quad SU_3 = -U_3.$$

It is easily seen that S is an isometric automorphism of (n, <,>) but now $S^4 = I$. Hence (N, <, >) is also a *4-symmetric space.*

So we have proved

THEOREM 9.9. *The six-dimensional group of type* H *is* 3- *and* 4-*symmetric.*

We note that these two facts are implicitly included in [32]. To show this we will now give another description of the six-dimensional example.

Let $N = \mathbb{R}^4 \times_{\varphi} \mathbb{R}^2$ denote the semi-direct product where the product is defined by

$$(a,b,c,d,r,s)(x,y,z,t,u,v) =$$

$$(x+a,y+b,z+c-rx+sy,t+d-sx-ry,u+r,v+s).$$

The metric

$$g = dx^2 + dy^2 + du^2 + dv^2 + (dz+udx-vdy)^2 + (dt+vdx+udy)^2$$

is a left invariant metric for this six-dimensional Lie group. Further the forms

$$\begin{cases} \theta_1 = dx , & \theta_4 = dt + vdx + udy , \\ \theta_2 = dy , & \theta_5 = du , \\ \theta_3 = dz + udx - vdy , & \theta_6 = dv , \end{cases} \tag{9.59}$$

and the dual vector fields

$$\begin{cases} E_1 = \frac{\partial}{\partial x} - u\frac{\partial}{\partial z} - v\frac{\partial}{\partial t} , & E_4 = \frac{\partial}{\partial t} , \\ E_2 = \frac{\partial}{\partial y} + v\frac{\partial}{\partial z} - u\frac{\partial}{\partial t} , & E_5 = \frac{\partial}{\partial u} , \end{cases} \tag{9.60}$$

$$E_3 = \frac{\partial}{\partial z}, \qquad E_6 = \frac{\partial}{\partial v}$$

are clearly left invariant. Hence (E_i, i = 1,...,6) determines the Lie algebra n of N. Proceeding in the same way as for the Heisenberg group one finds for the Lie brackets of the basic vector fields E_i :

$$\left\{\begin{array}{lll}
[E_1,E_2] = 0, & [E_1,E_3] = 0, & [E_1,E_4] = 0, \\
[E_1,E_5] = E_3, & [E_1,E_6] = E_4; & \\
[E_2,E_3] = 0, & [E_2,E_4] = 0, & [E_2,E_5] = E_4, \\
[E_2,E_6] = -E_3; & & \\
[E_3,E_4] = 0, & [E_3,E_5] = 0, & [E_3,E_6] = 0; \\
[E_4,E_5] = 0, & [E_4,E_6] = 0; & \\
[E_5,E_6] = 0. & &
\end{array}\right. \tag{9.61}$$

It is easy to see that this is isomorphic to the Lie algebra defined by (9.57).

This way of considering the six-dimensional example can also be used to show more directly that there does not exist a tensor field T such that $T_{XX} = 0$. To do this one uses the Cartan structure equations and compute the curvature 2-forms. After a long calculation one finds for the Ricci tensor ρ :

$$\left\{\begin{array}{l}
\rho_{11} = \rho_{22} = \rho_{55} = \rho_{66} = -1, \\
\rho_{33} = \rho_{44} = 1,
\end{array}\right.$$

the other components being zero.

Next suppose that N admits a naturally reductive homogeneous structure. Then there must exist a 3-form T on N. Put $\widetilde{\nabla} = \nabla - T$. Then the conditions

$$\tilde{\nabla}\rho = 0 \ ,$$

$$(\tilde{\nabla}_{E_1} T)_{E_1 E_2 E_4} = (\tilde{\nabla}_{E_1} T)_{E_1 E_2 E_3} = 0 \ ,$$

$$(\tilde{\nabla}_{E_5} T)_{E_5 E_6 E_3} = (\tilde{\nabla}_{E_5} T)_{E_5 E_6 E_4} = 0$$

imply that T is completely determined. One finds

$$T = -3(\theta_1 \wedge \theta_3 \wedge \theta_5 + \theta_1 \wedge \theta_4 \wedge \theta_6 - \theta_2 \wedge \theta_3 \wedge \theta_6 + \theta_2 \wedge \theta_4 \wedge \theta_5).$$

Now it is easy to see that $\tilde{\nabla}_{E_3} T \neq 0$ which provides a contradiction.

The example above is a special case of the following family of metrics on $\mathbb{R}^6$:

$$g = dx^2 + dy^2 + du^2 + dv^2 + (c + \alpha^2)(dz + udx - vdy)^2$$
$$+ \beta^2(dt + vdx + udy)^2 + 2\alpha\beta(dz + udx - vdy)(dt + vdx + udy) \ ,$$

where α, β and c are constants. Kaplan's example is obtained for $c = 1$, $\alpha = 0$, $\beta = 1$.

It is shown in [32] that all these manifolds are 4-symmetric spaces.

E. REMARKS

a. More detailed research concerning the relation between k-symmetric spaces and general groups of type H would be of some interest.

b. Another candidate for a manifold such that the geodesic symmetry is volume-preserving but which is not naturally reductive is a geodesic sphere in the Cayley plane. Indeed, all the geodesic spheres in Euclidean spaces and in the rank one symmetric spaces are homogeneous spaces but the Cayley plane is the only case where these spaces are not naturally reductive [58].

c. For an analytic Riemannian manifold the condition of being a D'Atri space is equivalent to an infinite series of conditions for the curvature tensor and its covariant derivatives. These are just the odd conditions which one obtaines when one expresses that a manifold is a harmonic space (see [5]). The first condition of this infinite series is the following condition for the Ricci tensor ρ :

$$\nabla_X \rho_{XX} = 0 \tag{9.62}$$

for all $X \in \mathfrak{X}(M)$.

Riemannian manifolds such that ρ satisfies (9.62) are studied in detail in [14]. In this context it would be of some interest to study all the homogeneous manifolds which satisfy (9.62) in relation to the homogeneous structures T on these manifolds.

For example it is proved in [48] that if the algebra of invariant differential operators of a Riemannian homogeneous space M is commutative, then M satisfies (9.62). In fact this property is much more general. Indeed it can be proved that when the algebra of differential operators on a Riemannian manifold is commutative, then the manifold is a commutative space and a commutative space is necessarily a D'Atri space. Hence we have (9.62). In fact we have an infinite series of conditions for the curvature tensor (see for example [51]).

10. SELF-DUAL AND ANTI-SELF-DUAL HOMOGENEOUS STRUCTURES

The study of the four-dimensional case is especially important in the theory of Riemannian manifolds. This is very well illustrated by the theory of *self-duality* and *anti-self-duality* (see for example [2]). This special feature of four-dimensional spaces is due to the fact that the rotation group SO(4) is not simple but locally isomorphic to SU(2) × SU(2). It is for this reason that the space of curvature tensors $\mathcal{R}(V)$ over a four-dimensional real vector space V with inner product has an extra decomposition. In general $\mathcal{R}(V)$ splits into three irreducible invariant subspaces under the action of the orthogonal group. But in dimension four the conformal invariant part $\mathcal{W}$ decomposes further into $\mathcal{W}_+ \oplus \mathcal{W}_-$ (see [45]). Let W denote the projection of a curvature tensor R on $\mathcal{W}$, i.e. W is the *Weyl tensor*. Then one defines : An oriented four-dimensional Riemannian manifold is *self-dual* (or *anti-self-dual* respectively) if its Weyl tensor $W = W_+$ (or $W = W_-$ respectively), i.e. if $W_- = 0$ (or $W_+ = 0$ respectively).

In this chapter we study the decomposition of the space $\mathcal{T}(V)$ under the action of the special orthogonal group if dim V = 4. In this case we first recall that

$$\mathcal{T}(V) = \mathcal{T}_1(V) \oplus \mathcal{T}_2(V) \oplus \mathcal{T}_3(V)$$

where

$$\dim \mathcal{T}_1(V) = 4 , \quad \dim \mathcal{T}_2(V) = 16 , \quad \dim \mathcal{T}_3(V) = 4.$$

We denote by Λ^2 the space of exterior 2-forms on V and note that

$$\mathcal{T}(V) = V^* \otimes \Lambda^2.$$

V is equipped with an inner product < , > and in what follows we fix an orientation on (V, <, >). The *Hodge star operator* $* : \Lambda^2 \to \Lambda^2$ is defined by

$$* \alpha \wedge \beta = (\alpha, \beta)\omega \in \Lambda^4$$

where $\alpha, \beta \in \Lambda^2$. (α,β) is the induced inner product of the 2-forms α, β and ω is the volume form defined by <, > and the orientation. Note that $*$ is a symmetric linear operator such that $*^2 = 1$. Then Λ^2 splits into a direct sum

$$\Lambda^2 = \Lambda^2_+ \oplus \Lambda^2_-$$

where $\Lambda^2_\pm$ are the ± 1 eigenspaces of $*$. A 2-form of Λ^2_+ (or Λ^2_- respectively) is called *self-dual* (or *anti-self-dual* respectively).

Since $\mathcal{T}_1(V) \cong V^*$ and $\mathcal{T}_3(V) \cong V^*$, these two spaces are also irreducible under the action of SO(4), but it follows from the general theory (see for example [53]) or from the remarks above that $\mathcal{T}_2(V)$ splits further into two irreducible components. Hence we have

THEOREM 10.1. *Let* V *be an oriented four-dimensional real vector space with inner product* < , > . *Then we have the orthogonal direct sum*

$$\mathcal{T}(V) = \mathcal{T}_1(V) \oplus \mathcal{T}_2^+(V) \oplus \mathcal{T}_2^-(V) \oplus \mathcal{T}_3(V) \tag{10.1}$$

where the summands are irreducible, invariant subspaces under the action of SO(4). *Further*

$$\dim \mathcal{T}_2^+(V) = \dim \mathcal{T}_2^-(V) = 8$$

and

$$\begin{cases} \mathcal{T}_2^+(V) = \{T \in \mathcal{T}_2(V) \mid T_{X*(YZ)} = T_{XYZ}\}, \\ \mathcal{T}_2^-(V) = \{T \in \mathcal{T}_2(V) \mid T_{X*(YZ)} = -T_{XYZ}\}, \quad X,Y,Z \in V. \end{cases} \tag{10.2}$$

The projections of $T \in \mathcal{T}(V)$ *are given by* (3.12), (3.13), (3.14), (3.15) *and*

$$\begin{cases} p_2^+(T)_{XYZ} = \frac{1}{2}\{p_2(T)_{XYZ} + p_2(T)_{X*(YZ)}\}, \\ p_2^-(T)_{XYZ} = \frac{1}{2}\{p_2(T)_{XYZ} - p_2(T)_{X*(YZ)}\}, \end{cases} \tag{10.3}$$

where $X,Y,Z \in V$.

This leads to

<u>DEFINITION 10.2.</u> Let T be a homogeneous structure on an oriented four-dimensional Riemannian manifold. Then T is said to be *self-dual* (or *anti-self-dual* respectively) if $p_2^-(T) = 0$ (or $p_2^+(T) = 0$ respectively).

We finish this chapter by giving an example of a manifold with an anti-self-dual homogeneous structure. In chapter 8 we noted that any four-dimensional 3-symmetric space is almost Kählerian with respect to any invariant almost Hermitian structure. Therefore let (M,g) be a 3-symmetric space with dim M = 4 and denote by J the canonical almost complex structure. Consider the orientation given by J and the homogeneous structure T determined by

$$T_X = \frac{1}{2} J\nabla_X J, \qquad X \in \mathfrak{X}(M). \tag{10.4}$$

It follows from Theorem 8.1 that $T \in \mathcal{T}_2$. Further, using a basis (e_1,e_2,Je_1,Je_2) at each point $m \in M$, it is not difficult to prove that

$$p_2^+(T) = 0.$$

Hence we have

<u>EXAMPLE 10.3.</u> *Let* (M,g) *be a four-dimensional 3-symmetric space with the orientation determined by the canonical almost complex structure on* M. *Then the homogeneous structures* T *given by* (10.4) *is anti-self-dual.*

Note that a change of this orientation provides a self-dual homogeneous structure.

REFERENCES

[1] Ambrose, W. & Singer, I.M. (1958). On homogeneous Riemannian manifolds. Duke Math. J., 25, 647-669.

[2] Atiyah, M., Hitchin, N. & Singer, I.M. (1978). Self-duality in four-dimensional Riemannian geometry. Proc. Roy. Soc. London, A 362, 425-461.

[3] Azencott, R. & Wilson, E.N. (1976). Homogeneous manifolds with negative curvature, Part I. Trans. Amer. Math. Soc., 215, 323-362.

[4] Azencott, R. & Wilson, E.N. (1976). Homogeneous manifolds with negative curvature, Part II. Mem. Amer. Math. Soc., 178.

[5] Bérard Bergery, L. (1981). Les espaces homogènes riemanniens de dimension 4. In Géométrie riemannienne en dimension 4, pp. 40-60. Séminaire A. Besse. Paris : Cedic.

[6] Berger, M., Gauduchon, P. & Mazet, E. (1971). Le spectre d'une variété riemannienne. Lecture Notes in Mathematics, 194. Berlin, Heidelberg, New York : Springer-Verlag.

[7] Brown, R. & Gray, A. (1967). Vector cross products. Comment. Math. Helv., 42, 222-236.

[8] Cheeger, J. & Ebin, D.G. (1975). Comparison theorems in Riemannian geometry. Amsterdam : North-Holland Publ. Co..

[9] D'Atri, J.E. (1975). Geodesic spheres and symmetries in naturally reductive homogeneous spaces. Michigan Math. J., 22, 71-76.

[10] D'Atri, J.E. & Nickerson, H.K. (1969). Divergence-preserving geodesic symmetries. J. Differential Geometry, 3, 467-476.

[11] D'Atri, J.E. & Nickerson, H.K. (1974). Geodesic symmetries in spaces with special curvature tensor, J. Differential Geometry, 9, 251-262.

[12] D'Atri, J.E. & Ziller, W. (1979). Naturally reductive metrics and

Einstein metrics on compact Lie groups. Mem. Amer. Math. Soc., 215.

[13] Gray, A. (1972). Riemannian manifolds with geodesic symmetries of order 3. J. Differential Geometry, 7, 343-369.

[14] Gray, A. (1978). Einstein-like manifolds which are not Einstein. Geometriae Dedicata, 7, 259-280.

[15] Gray, A. & Hervella, L.M. (1980). The sixteen classes of almost Hermitian manifolds and their linear invariants. Ann. Math. Pura Appl., 123, 35-58.

[16] Gray, A. & Vanhecke, L. (1979). Riemannian geometry as determined by the volume of small geodesic balls. Acta Math., 142, 157-198.

[17] Heintze, E. (1974). On homogeneous manifolds of negative curvature. Math. Ann., 211, 23-34.

[18] Helgason, S. (1968). Lie groups and symmetric spaces. In Batelle Rencontres 1967. Lectures in Mathematics and Physics, pp. 1-71. New-York : Benjamin.

[19] Helgason, S. (1978). Differential geometry, Lie groups and symmetric spaces. New York : Academic Press.

[20] Jensen, G.R. (1969). Homogeneous Einstein spaces of dimension four. J. Differential Geometry, 3, 309-349.

[21] Kaplan, A. (1980). Fundamental solutions for a class of hypoelliptic PDE generated by composition of quadratic forms. Trans. Amer. Math. Soc., 258, 147-153.

[22] Kaplan, A. (1981). Riemannian nilmanifolds attached to Clifford modules. Geometriae Dedicata, 11, 127-136.

[23] Kaplan, A. On the geometry of groups of Heisenberg type. To appear in Bull. Lond. Math. Soc..

[24] Kiričenko, V.I. (1980). On homogeneous Riemannian spaces with invariant tensor structure. Soviet Math. Dokl., 21, 734-737.

[25] Kobayashi, S. (1972). Transformation groups in differential geometry. Ergebnisse der Mathematik und ihrer Grenzgebiete, 70. Berlin, Heidelberg, New York : Springer-Verlag.

[26] Kobayashi, S. & Nomizu, K. (1963 and 1969). Foundations of differential geometry, I and II. New York : Interscience Publishers.

[27] Kostant, B. (1955). Holonomy of the Lie algebra of infinitesimal motions of a Riemannian manifold. Trans. Amer. Math. Soc., 80, 528-542.

[28] Kostant, B. (1960). A characterization of invariant affine connections. Nagoya Math. J., 16, 35-50.

[29] Kowalski, O. (1975). Classification of generalized symmetric Riemannian spaces of dimension n ⩽ 5. Praha : Rozpravy ČSAV, Řada MPV, no 8, 85.

[30] Kowalski, O. (1980). Generalized symmetric spaces. Lecture Notes in Mathematics, 805. Berlin, Heidelberg, New York : Springer-Verlag.

[31] Kowalski, O. (1980). Additive volume invariants of Riemannian manifolds. Acta Math., 145, 205-225.

[32] Kurcius, E. (1978). 6-dimensional generalized symmetric Riemannian spaces. Ph. D. Thesis, University of Katowice.

[33] Milnor, J. (1976). Curvature of left invariant metrics on Lie groups. Advances in Math., 21, 293-329.

[34] Miquel, V. (1982). The volumes of small geodesic balls for a metric connection. Compositio Math., 46, 121-132.

[35] Miquel, V. & Naveira, A.M. (1980). Sur la relation entre la fonction volume de certaines boules géodésiques et la géométrie d'une variété riemannienne. C.R. Acad. Sc. Paris, Série A, 290, 379-381.

[36] Nomizu, K. (1954). Invariant affine connections on homogeneous spaces. Amer. J. Math., 76, 33-65.

[37] Nomizu, K. (1960). On local and global existence of Killing vector fields. Ann. of Math., 72, 105-120.

[38] Nomizu, K. (1962). Sur les algèbres de Lie de générateurs de Killing et l'homogénéité d'une variété riemannienne. Osaka J. Math., 14, 45-51.

[39] Palais, R. (1957). A global formulation of the Lie theory of transformation groups. Mem. Amer. Math. Soc., 22.

[40] Pansu, P. (1982). Géométrie du groupe de Heisenberg. Thèse 3ème cycle, Université Paris VII.

[41] Postnikov, M.M. (1967). The variational theory of geodesics. Translated from the Russian. Philadelphia, London : W.B. Saunders Co.

[42] Sato, T. (1979). Riemannian 3-symmetric spaces and homogeneous K-spaces. Mem. Fac. of Technology Kanazawa Univ., 12(2), 137-143.

[43] Sekigawa, K. (1978). Notes on homogeneous almost Hermitian manifolds, Hokkaido Math. J., 7, 206-213.

[44] Singer, I.M. (1960). Infinitesimally homogeneous spaces. Comm. Pura Appl. Math., 13, 685-697.

[45] Singer, I.M. & Thorpe, J.A. (1969). The curvature of 4-dimensional Einstein spaces. In Global Analysis Papers in Honor of K. Kodaira, eds. D.C. Spencer & S. Iyanaga, pp. 355-365. Princeton : Princeton University Press & University of Tokyo Press.

[46] Spanier, E.H. (1966). Algebraic Topology.. New York : McGraw-Hill.

[47] Steenrod, N. (1951). The topology of fibre bundles. Princeton : Princeton University Press.

[48] Sumitomo, T. (1972). On a certain class of Riemannian homogeneous spaces. Colloquium Math., 26, 129-133.

[49] Tricerri, F. & Vanhecke, L. (1981). Curvature tensors on almost Hermitian manifolds. Trans. Amer. Math. Soc., 267, 365-398.

[50] Tricerri, F. & Vanhecke, L. Naturally reductive homogeneous spaces and generalized Heisenberg groups. To appear.

[51] Vanhecke, L. Some solved and unsolved problems about harmonic and commutative spaces. Bull. Soc. Math. Belg. B 34, 1-24, 1982.

[52] Vanhecke, L. & Willmore, T.J. Interaction of tubes and spheres. To appear in Math. Ann.

[53] Weyl, H. (1946). Classical groups, their invariants and representations. Princeton : Princeton University Press.

[54] Wilson, E.N. (1982). Isometry groups on homogeneous nilmanifolds. Geometriae Dedicata, 12, 337-346.

[55] Wolf, J.E. (1964). Homogeneity and bounded isometries in manifolds of negative curvature. Illinois J. Math., 8, 14-18.

[56] Wolf, J.E. (1967). Spaces of constant curvature. New York : McGraw-Hill.
(1977) Spaces of constant curvature. Berkeley : Publish or Perish.

[57] Wolf, J.E. (1968). The geometry and structure of isotropy irreducible homogeneous spaces. Acta Math., 120, 59-148.

[58] Ziller, W. (1982). Homogeneous Einstein metrics on spheres and projective spaces. Math. Ann., 259, 351-358.

INDEX